DRAFTING ENGINEERING CONTRACTS

Drafting
Engineering Contracts

HENRY HENKIN
C.Eng., F.I.C.E., M.I.Struct.E.,
London, UK

Routledge
Taylor & Francis Group

LONDON AND NEW YORK

First published 1988 by Taylor & Francis

Published 2020 by Routledge
2 Park Square, Milton Park, Abingdon, Oxon OX14 4RN
52 Vanderbilt Avenue, New York, NY 10017, USA

First issued in paperback 2020

Routledge is an imprint of the Taylor & Francis Group, an informa business

British Library Cataloguing in Publication Data

Henkin, H.
  Drafting engineering contracts.
  1. Great Britain. Engineering. Contracts
  I. Title
  620

Library of Congress Cataloging in Publication Data

Henkin, H. (Henry)
  Drafting engineering contracts by H. Henkin.
    p.   cm.
  ISBN 1-85166-223-5
  1. Engineering—Contracts and specifications—Great Britain.
  2. Engineering—Contracts and specifications.   I. Title.
  KD1641.H46 1988
  343.4′078624—dc 19
  [344.10378624]   88-7287

  ISBN 1-85166-223-5

Publisher's Note
The publisher has gone to great lengths to ensure the quality of this reprint
but points out that some imperfections in the original may be apparent

ISBN 13: 978-0-367-58025-4 (pbk)
ISBN 13: 978-1-85166-223-4 (hbk)

Foreword

by

P. A. COX B.Sc(Eng)., F.Eng., F.I.C.E.
Past President of the Institution of Civil Engineers
Lately Chairman of Rendel Palmer & Tritton Ltd

The drafting of Engineering Contracts is often regarded as being primarily concerned with that part of Contract Documents known as the Conditions of Contract. Those who have been responsible for the control of contracts will recall vividly that many problems and claims arose from other parts of the Documents and also the relationship between the parts. The integrity of the whole of the Contract Documents is therefore the hallmark of good drafting and essential to successful project implementation.

Henry Henkin has had extensive experience in the United Kingdom and overseas in the drafting of Contract Documents and in the administration of a wide range of contracts, both as a contractor and as a supervising engineer. That experience has enabled him to set down in this volume a clear methodology for drafting. Requirements are well reasoned and the many examples give the reader a clear understanding of the need to control carefully the power available in our language. During his service with Rendel Palmer & Tritton, Henry showed his command of the subject, earning the appreciative respect of his colleagues, our clients and the contractors whose work we directed.

Contract forms – FIDIC, EDF, BPF and others – are continuing to be developed as client requirements evolve in the changing, and increasingly internationalised scene. This volume will be a valuable guide to those charged with that development as well as to those responsible for the detailed preparation of documents for specific projects.

Preface

When I first read Abrahamson's "Engineering Law and the ICE Contracts" I noted, in passing, the points he made concerning drafting and the books which he recommended on legal drafting, but I must admit that I didn't read those books until I came to write this book. I found that the books were difficult to obtain and no engineers of my acquaintance appeared to have read them.

In the late seventies, when I already had considerable experience of drafting documents, I had to draft new civil engineering Conditions of Contract for a Commonwealth government whose existing Conditions were an out of date colonial form which did not suit the current contracts entered into by their Ministries and other national organisations; I also had to oversee the preparation of various mechanical and electrical engineering standard forms. To do this I found it necessary to analyse the basic requirements for the drafting of lengthy documents and, in particular, to set out the logical framework for Conditions of Contract and for other documents. I realised then that there was no readily accessible publication which would assist me in this. There lies the origins of this book, which is intended to present the ideas and methods of the authorities on legal drafting in the context of engineering contract practice, together with detailed discussion of the requirements for form and content of the documentation which engineers are required to draft.

The readers of this book may vary from young graduates putting together their first specification for a small work to experienced engineers drafting major documents, but who have not had any formal training in drafting; they may include civil, chemical, structural, mechanical and electrical engineers. I have not, therefore, expected that (apart from the necessary engineering knowledge) my readers would necessarily know more than the elements of contract law and have some limited experience of contract administration, although much greater knowledge and experience is required for drafting important documents.

There are very few references to legal judgements in the book and then only in connection with interpretation and punctuation. As the book is not primarily concerned with law, I have considered it inappropriate to give case references in those few instances.

Although the book is concerned with drafting and not with law or administration nor with the subject matter of individual contracts, these various matters cannot be ignored. I have tried to overcome this problem by the introduction of a variety of examples and by setting out a large number of the items which need to be included in the various documents, so that the text may serve also as an aide-mémoire. The book is based upon British contract practices, but the principles described are applicable wherever English is used as the language of documents.

I consider that the provisions concerning payments to the Contractor (and contra charges against him) are inadequately and illogically dealt with in the standard forms of Conditions. In particular, I consider it essential that *all* entitlements to payments be brought together in one Section and that references to them should, so far as possible, be eliminated from other parts of the contract documents. I have tried to emphasise this in the text and I hope that my comments may lead to some discussion and improvement on this important subject.

As a civil engineer my examples of drafting relate, in many cases, to civil engineering and construction contracts in which I have been involved in the past 25 years. As I have also been responsible for drafting mechanical and electrical engineering documents and for assisting and redrafting documents prepared by engineers in other disciplines, I have been able to include some material relating specifically to mechanical, electrical and chemical engineering. However, the principles involved in the drafting of contract documents apply to all types of engineering work and I have, therefore, tried to concentrate on those principles while relying on examples to demonstrate their implementation.

Civil engineers use Bills of Quantities extensively for arriving at the contract price by remeasurement. There would be little point in adding one further text to the many publications concerning Bills of Quantities except to the extent necessary to provide explanations to engineers in other disciplines who are not familiar with these documents, but who may find it useful to understand the principles involved. There appears to be very little published concerning the details of payment schedules required for lump sum and other types of contract and I hope that my explanations and examples on Payment Documents will assist those engineers who have to prepare contracts which are not remeasured and for which Bills of Quantities are not applicable.

Finally, I hope that all my readers will be allowed adequate time for drafting and checking their documents. Unfortunately, in most cases this will not be so. This appears to be a major problem in modern engineering practice.

Acknowledgements

My thanks to Mr. P. A. Cox for his foreword which is highly appreciated and valued, as well as for his encouragement and assistance in enabling me to make use of documents drafted for Rendel Palmer & Tritton as a basis for many of the examples in the text. The wording of those examples has, of course, been changed from that in the original text and the responsibility for those examples is entirely mine and is not attributable to Rendel Palmer & Tritton.

My thanks also to Mr. John Chandler, M.A., C.Eng., F.I.C.E., F.C.I.Arb., and to Mr. Fred McAdam, B.Sc., D.I.C., D.M.S., C.Eng., F.I.C.E., M.I.W.E.M., F.B.I.M. for reading through the manuscript and giving me their very valuable comments based on their extensive experience of engineering contracts.

My wife typed the whole of the manuscript from my dictation and I have this and much else to thank her for.

Acknowledgements

Terminology and Abbreviations

(1) The term Employer is used throughout this book to describe the party to the contract who employs the Contractor to carry out the Works, although in mechanical and electrical contracts he is more usually referred to as the Purchaser and in government contracts he is referred to as the Authority. Similarly, the term Engineer is used to describe the person nominated to administer the contract, although in government contracts he is referred to as the Supervising Officer (SO) and in building contracts he is usually replaced by the Architect. In some cases the Company Representative is substituted for the Engineer but, as explained in Chapter 4 Part 4.1, this does not seriously affect the drafting.

(2) The term draftsman is intended to describe anyone (male or female) responsible for preparing contract documents. It is not to be confused with those who prepare drawings, previously called draughtsmen but now referred to as technicians.

(3) Each Chapter has been divided (and where necessary sub-divided) using decimal notation. The divisions and sub-divisions are referred to as Parts in order to avoid possible confusion with reference to Sections of contract documents.

(4) The following abbreviations are used; except for the FIDIC Conditions, the documents are all published in the UK, but many similar documents are available in other countries where English type legal systems prevail, although there are differences due to the effect of local laws:–

Abbreviations	Full Title
Conditions	Conditions of Contract or General Conditions of Contract.
ICE Conditions	Conditions of Contract and Forms of Tender, Agreement and Bond for use in Connection with Works of Civil Engineering Construction.
I.E.E. Conditions	Model Form of General Conditions of Contract including forms of Agreement and Guarantee recommended by the Institution of Mechanical Engineers, the Institution of Electrical Engineers and the Association of Consulting Engineers for use in connection with Home Contracts – With Erection. (Model Form A/1976).

Abbreviations	*Full Title*
I.Chem.E. Conditions	Model Form of Conditions of Contract for Process Plants Suitable for Lump-sum contracts in the United Kingdom.
FIDIC Civil Conditions	Conditions of Contract (International) for Works of Civil Engineering Construction with forms of tender and agreement (3rd Edition 1977).
FIDIC E & M Conditions	Conditions of Contract (International) for Electrical and Mechanical Works (including erection on site) with Forms of Tender and Agreement (2nd Edition 1980).
GC/Works/1	General Conditions of Government Contracts for Building and Civil Engineering Works.
JCT Conditions	'Standard Form' of building contract for use with quantities private edition.
Roads & Bridges Specification	Specification for Road and Bridge Works. (This has now been superseded by the Highway Works Specification).
Highway Works Specification	Specification for Highway Works.

Note: The names of the publishers of these documents and the latest edition is given in the bibliography.

Contents

CHAPTER 1 — GENERAL PRINCIPLES

CHAPTER 2 — WORDS, SENTENCES AND PUNCTUATION

CHAPTER 3 — ARRANGEMENT AND FORM OF DOCUMENTS

CHAPTER 4 — CONDITIONS OF CONTRACT

CHAPTER 1

General Principles

1.1 Introduction

A large Japanese contractor has said that he has "completed work on a 50 storey office block at a cost of £29 m without any contract". His reference was to lack of a formal contract; there must, of course, have been a verbal contract in which the parties agreed that the work would be carried out for a particular price or based upon a particular method of pricing and there must have been drawings and a specification describing the work to be done. There may even have been a pricing document, or access to the contractor's estimate, for the purpose of pricing variations.

In theory, engineering or building contracts can consist solely of a simple agreement in which the Contractor agrees to carry out the work for a fixed sum. In practice the complexity of the work demands drawings and documentation, usually of considerable length, to supplement the simple contract and to provide the rules governing the carrying out of the work and the relations between the parties.

Provided that the legal requirements for a valid contract are observed, a formal agreement is not required; many commercial contracts for engineering work are in the form of a simple exchange of letters, but they are backed by formal documentation which, in most English speaking countries, usually comprises:–

Conditions of Contract
Specification
Drawings

Bills of Quantities or Schedules of Rates or some other Payment Document.

Contracts with public authorities and, usually, with large commercial companies, will have a simple formal agreement which incorporates a formal tender by the Contractor. The formal agreement is usually (as in the ICE Agreement form) a simple listing of the contract documents and a legal agreement to abide by them.

The form of the basic documents may vary widely. For example, some engineers consider the Specification to be an extension of the Conditions of Contract and draft it accordingly, numbering the clauses in continuation of the clause numbers in the Conditions. In building contracts, on the other hand, the Specification usually forms part of the Bills of Quantities, the theoretical intention being that tenderers should have the opportunity of pricing the various items of the Specification (in practice this does not usually occur). Bills of Quantities or Schedules of Rates may be absent or may vary widely in form according to the method of measurement; in the case of a lump sum contract there may be merely a schedule of interim payments, although it is common to include a schedule of rates to enable variations to be easily priced. The above list of basic documents is not exhaustive; other documents such as drainage schedules and earthwork schedules, performance bonds and retention bonds, supplementary agreements with outside parties, nominated sub-contract forms, quotations for pre-ordered materials, forms of tender, instructions for tendering, etc, are common. Package deal contracts and management contracts may have different documentation, although, in the absence of nationally agreed forms specific to package deal or management contracts, there is often a tendency to incorporate standard documents used in other types of contracts. This documention is in addition to the large number of drawings which usually form part of the contract. The wide variety and type of documentation, the variety of site and manufacturing conditions and the variety of financial arrangements makes each engineering contract unique, requiring careful consideration and preparation of the documentation and modification of those standard documents which are incorporated in it.

A large engineering contract may have between 500 and 1500 pages of written documentation in addition to up to perhaps 2000 or more drawings. Apart from the drafting of bonds and similar documents and the checking of standard documentation, the whole of this very large and complex legal document is usually drafted by engineers, with varying degrees of skill. The preparation of the various contract documents requires a substantial knowledge of the design and construction of engineering works and of the manner in which they are supervised and administered. With the exception of those rare individuals who qualified and worked as engineers before becoming lawyers, it is not practical for lawyers to acquire this knowledge in order to draft engineering documents, although some specialist lawyers have acquired sufficient knowledge of the administration of the contracts and of employer–contractor–engineer relationships to be able to advise on contract conditions and to take instructions and draft Conditions of Contract. There is really no satisfactory alternative to engineers acquiring the necessary skills in legal drafting to enable them to prepare sound contract documents.

Drafting skills have generally been acquired by 'apprenticeship', the young engineer copying previous documentation with minor modifications and gradually acquiring the prevailing style and thereafter proceeding to draft documents which require more original work than is forthcoming in available or standard documentation. For historical reasons, legal documents were, in the past, lengthy, complex in construction and verbose, written in a legal jargon which was often incomprehensible to the clients for whom the documents had been prepared. In the past century the legal profession and legislation has been moving toward clear and less verbose documents, less complex in verbal construction and easier for the layman to read, understand, and administer.

There have been many legal volumes dealing with the drafting of particular types of document, based upon standard forms produced by their authors or generally in current use, but very few dealing solely with the general and detailed requirements for legal drafting. However, of recent years there appears to have been moves in the legal community to provide lawyers with education in legal drafting to improve and supplement the knowledge acquired by 'apprenticeship'. A number of books have appeared dealing solely or mainly with the general and detailed requirements for legal drafting (as distinct from the requirements for individual documents). The stress among these and earlier authorities is on clear and straightforward writing with the minimum of legal jargon, so as to make the documents intelligible to all who have to use them.

This book deals with all those aspects of writing, style, form and presentation required in the drafting of engineering contract documents. It is only incidentally concerned with the substance of the matters with which those documents deal. In consequence it deals only incidentally with the subject of drawings, although they form part of the documentation. Many examples are given to illustrate the text, but those examples are not necessarily suited to all contracts and should not be slavishly copied.

1.2 Knowing The Subject

To draft legal documents one must obviously have some knowledge of law. The young engineer commencing to draft part of a large specification (usually based on a specification for a previous job) requires only a very rudimentary knowledge, but the more senior engineer responsible for preparing the whole of the documentation for a contract ought to have a sound knowledge of construction law. The engineer cannot be expected to have the knowledge of a lawyer or the outlook and background which comes from the full time study and practice of law. He would not expect to be able to quote precedents, nor to delve into related subjects such as insurance law and practice or company law, which affects some aspects of engineering contracts. He must expect to have a sufficient knowledge to recognise legal problems and to seek advice on matters outside his experience. It is more important to have a good knowledge of the standard forms of Conditions of Contract and of the legal implications of the various clauses in them than to have a high standard of theoretical knowledge of the law of contracts.

In legal theory, an independent contractor who has contracted to carry out work is free to carry it out in the manner he considers best. In engineering contracts this freedom of action is severely circumscribed; limitations are placed upon the contractor so as to ensure that:

(i)   his actions do not make the Employer vicariously liable in respect of special and general legal obligations;

(ii)   the Contractor has a management control system which will ensure that the work is properly carried out (in the Employer's interest) and not merely for the purpose of maximising profit (in the Contractor's interest);

(iii)   the Contractor demonstrates that his construction methods and materials are satisfactory for the work;

(iv)   methods of construction are those current in the industry or which can be proved, by demonstration, to be satisfactory in the circumstances of the site and the work;

(v) if possible, the Employer is provided with sanctions within the Contract (without prior legal action) if the Contractor fails to fulfil his responsibilities.

All the documents have to take account of the foregoing requirements; engineering knowledge and experience is required to deal with all but the first of these requirements and such knowledge is, therefore, an essential part of the qualifications for drafting engineering contracts. These requirements are particularly applicable to engineering construction contracts because:

(i) in many cases examination of the finished work will not reveal whether the work has been properly carried out;

(ii) the incorporation of unacceptable materials or workmanship may prove irreversible without substantial demolition and considerable delay or may even prove impracticable due to site conditions;

(iii) the Employer usually wishes to minimise his own involvement in the construction process and to minimise additional costs.

The requirements outlined apply also, to a greater or lesser extent, to contracts involving manufacturing and erection of engineering work and, to a lesser extent, to the manufacture of specially designed engineering structures or machines.

It is important for the draftsman to know the requirements of his client, whether that client be a public authority instructing a consulting engineer or the estimating director of a contractor offering a package deal contract. Client's instructions fall into two parts:–

(1) Direct instructions concerning matters such as the form of Conditions of Contract and any amendments thereto, methods of calculating liquidated damages, special requirements to be incorporated in the Specification and in the Payment Documents, requirements concerning performance bonds, and the like.

(2) Instructions concerning the design of the Works including client's approval of drawings, of draft documents and of tender particulars (such as contract period, etc.).

The draftsman must work closely with the designer to ensure that the materials and methods of construction required by the design are properly incorporated in the documents and that there is proper cross-referencing between the documents and the drawings; all too often it is found that drawings and specifications are contradictory because of lack of such liaison. Although, in theory, the designer and the document draftsman might be the same person, in practice they are often separate individuals. Because of the importance of design liaison and involvement in the preparation of contract documents, it is essential that the document draftsman commence work at an early stage in the design process. This should preferably be when the preliminary design has been agreed with the client and instructions have been given to commence detailed design.

The last, but by no means the least, requirement for document drafting is the ability to write clear and unambiguous English. This requires the avoidance of unnecessary jargon and a knowledge of modern English usage, rather than familiarity with pedagogic grammatical rules.

1.3 Intention of Parties to a Contract

The purpose of contract documents is to express and record unambiguously the matters intended by the parties to the contract. However, that intention is expressed by the words of the contract and not by what the parties supposed or claim as their intention. If the intention to be adduced from the contract wording does not correspond to what either of the parties believe was intended, then that belief is of no avail, for the contract will be interpreted in accordance with the wording. In a judgement in 1861 Lord Wensleydale said –

"The question is not what the parties to a deed or other documents may have intended to do by entering into that deed, but what is the meaning of the words used in that deed: a most important distinction in all cases of construction and disregard of which often leads to erroneous conclusions".

It was even more strongly put in the judgement by Sir Gorell Barnes in 1907 when he said –

"What a man intends and the expression of his intention are two different things. He is bound and those who take after him are bound by his expressed intention. If that expressed intention is unfortunately different from what he really desires, so much the worse for those who wish the actual intention to prevail."

These judgements underline the need for clear accurate drafting to express correctly and unambiguously the intention of the contract. To ensure that the intention is correctly expressed, it is necessary to consider all the alternative possibilities that might arise from the wording being used, and so to guard against unforeseen interpretations. Generally, engineering documents tend to be lengthy and usually have to be drafted to a tight time schedule: it is often impossible to find the time to examine every possibility on all the clauses. It is common, therefore, to rely to a considerable extent on standard forms which have been carefully thought out and used over a number of years or on clauses from previous documents which appear to have proved satisfactory. Unfortunately, clauses from previous documents do not always prove adequate for the other work and there is really no substitute for experience and care. Drafting is an art which can only be acquired with practice, but a little study of the principles involved will undoubtedly help.

1.4 Basic Requirements for Document Drafting

Writers on legal drafting have summarised the methodology of drafting; the following five basic precepts, based on that methodology, provide good general guidance in the drafting of engineering contracts.

A. Before work commences on drafting the wording of the document, the whole layout and overall design should be conceived as a framework for drafting. It is worthwhile preparing a drafting plan in the form of a contents list divided into section headings and clause headings. During drafting it may be necessary to change or add to or rearrange the clauses in the original contents list, but such changes should concern matters of detail. The basic content and outline should be retained; if changes of principle are to be made then the outline and content should be redrawn rather than amended, even if a substantial part of the document has been drafted. Obviously, if an engineering structure is to be steel framed instead of concrete framed, there has to be a radical change of plan. Even the clauses concerning the concrete will have to be reconsidered and rearranged; they cannot merely be transferred as a whole from one specification to the other. Unfortunately, in the case of Conditions of Contract this precept is more often honoured in the breach; attempts are made to adopt standard forms (often written for conventional measure and value contracts) with minor modifications to suit radically different concepts, such as design and construct contracts or contracts involving a degree of consultancy, such as those for site investigation.

B. A logical order should be adopted for the various sections and clauses. The actual logic applicable will depend upon the nature of the document and the nature of the work as well as the personal opinions of the draftsman. The order appropriate to various documents is discussed in later chapters. The precise order adopted by the draftsman is not of importance so long as it is logical and the logic of the arrangement is clear to the reader so that he is able to understand particular requirements in relation to the general requirements for the work.

C. Nothing should be included or omitted at random. If, during drafting, the draftsman cannot appreciate the relevance of some provision which it has become common to include or which has been included in a similar previous document, then he must either satisfy himself by further inquiries that the provision is necessary or he should omit it. Unnecessary repetition is also an irrelevance; where definitions are given in the document, repetition of part of the definition may be a dangerous irrelevance. As an example, the FIDIC Civil Conditions define 'approved' as meaning 'approved in writing'; to continually refer to approval 'in writing' in the Specification amounts to depreciating the definition and suggesting that there may be occasions when verbal approval is acceptable.

D. The documents should be in a form with which the reader will be familiar and the technical language should be that which is ordinarily used in the class of work. This does not preclude change of form to suit new types of work or new concepts, but it does assume that the draftsman will, in such cases, so far as possible retain familiar concepts and technical terms.

E. The language used should be precise and accurate, so that every phrase has a clear meaning and phrases are interconnected in a manner which will not give rise to ambiguity. It should be appreciated that clarity may be achieved equally by the omission of irrelevant phrases as by the use of apt wording.

1.5 Communication and Drafting

An engineering contract is only part of the process of construction of a project or of fabrication of some elements of it. The purpose of the contract documents is not only to set out the legal rules which are to govern the parties contractual conduct but also to communicate the requirements for carrying out the work and the stipulations and limitations applicable to materials and workmanship. The drafting of the documents is not, therefore, only a matter between the draftsman and his client, but has to be understood by a wide range of people who have to construct the work, or to receive instructions in connection with it, or to claim payment for it, or to provide insurance, or to carry out any of the many other functions involved. To ensure that the many people concerned correctly interpret the requirements, the draftsman must follow the principles of communication applicable to all written statements.

Reed Dickerson (in The Fundamentals of Legal Drafting) postulates four main elements in the written communication process –

(1) The author.
(2) The audience.
(3) The written utterance.
(4) The relevant context or environment.

The fourth element is, perhaps, the most misunderstood, even though all writers are constrained by it. Many engineers are misled by the legal requirement that the meaning of a contract is to be inferred only from the wording of the documents; they overlook the fact that the meaning of the words themselves depends upon the context in which the document is written and also that there are rules of law which introduce implied terms to ensure that certain types of contract conform with the applicable context or environment. The best known of these rules deals with contracts of sale and infers that the contract provides that the goods being sold are of merchantable quality or that a ship being sold is seaworthy, etc.

All communication is based upon the language habits of the particular groups being addressed. Meaning in language is based upon usage and, although small changes to usage are generally acceptable or may even improve intelligibility, substantial departure from accepted usage seriously detracts from communication. In general, adherence to the conventions of language currently adopted by the audience cannot be over emphasised; for example if carpentry is specified in the language of the soils engineer the audience for that communication might well be baffled.

1.6 The Author and The Audience

In most writing, including a large proportion of legal documents, the author is a clearly recognisable individual. Where a book or a report is written by a number of individuals, the responsibility for proper communication will be shared only by the limited number of individual authors concerned. However, in the case of the documents for a large and complex engineering construction contract, we find that, in practice, there is a much more complex form of authorship which imposes both limitations and external environment on the documents.

The Conditions of Contract are almost invariably standard forms which have usually been drafted by a committee and revised, over the years, by other committees. There are often additional clauses used by the Client as standard (dealing with such matters as taxation and legal responsibilities) and usually drafted by the Client's lawyers. There may also be clauses concerning relationships with other bodies, such as Statutory Undertakers, which arise from negotiations with those bodies. Amendments and additional clauses appropriate to the work will also be provided by the engineer responsible for drafting the documents, although in many large consulting offices these clauses and amendments may be drafted by a special contracts department. The result is generally a complicated patchwork of requirements and responsibilities; it usually takes some years of experience before an engineer is able to master the intricacies of the Conditions of Contract.

Bills of Quantities in the UK are based upon a standard Method of Measurement prepared by a committee, although in this case a consultant or a small drafting committee is likely to have co-ordinated the drafting and prepared the final draft as a consistent whole. Preambles to the Bills will have to be prepared by the measurement engineer or quantity surveyor, including modifications to the Method of Measurement. The Bill descriptions are often drawn from a standard library of descriptions which may have been computerised. The measurement engineer or quantity surveyor may not be responsible for the remainder of the documents, i.e. they may represent an additional author.

The main draftsman responsible for preparing the documents will usually draft the Specification, but the clauses concerning workmanship and materials are likely to be based on a standard specification or on clauses previously used for other work and suitably modified. The draftsman also prepares other documents such as the Instructions for

Tendering and coordinates all the various parts of the contract documents, including a standard form of tender and standard forms of agreement and bond.

Obviously, the draftsman responsible for preparing documents requires not only a knowledge of the engineering work involved and of the designer's intentions (if he is not also the designer) but also requires a knowledge of the industry and of the language habits of the various groups who have been involved in drafting standard documents.

The audience who must read and understand the documents are equally varied. The draftsman's first audience will be the client, who has to be satisfied that the documents express his requirements for the work. Large and complex engineering construction works are usually ordered by public authorities or large companies who have engineering and legal departments and who will usually scrutinise the documents and be familiar with the language habits of those who will have to read them. Such a client will not necessarily be familiar with all the specialities, but this will require explanation rather than adjustment of the documentation.

When the documents have been completed and approved they will usually be sent to a number of contractors who will be invited to tender for the work. There they will be dealt with by planners (who will usually be engineers with wide experience who will decide the working methods to be assumed for tender pricing) and by estimators who will be responsible for pricing the tender.

Both planners and estimators will usually be familiar with the language habits of the various audiences for the parts of the documents with which they are concerned, but the estimator will usually obtain quotations for materials from the contractor's buying department, who will think in terms of the language habits and jargon of the suppliers. The specification for materials must, therefore, take account of the language habits and jargon employed in the sale and distribution of the various materials. The estimator may also have to take account of the advice of the company's lawyers concerning loopholes in the Conditions of Contract and will also have to obtain quotations from insurance companies and/ or brokers for insurances and bonds which may be required.

In theory, the draftsman should introduce amendments to standard Conditions to close loopholes arising from known ambiguities or from the inapplicability of particular conditions to the work in hand; the draftsman will, however, find that contractors' lawyers are usually one jump ahead. The various standard Conditions of Contract have insurance clauses drafted in the terminology of that industry; in most cases it is sufficient for the draftsman to have a general understanding of the terminology, in order to avoid inserting in other parts of the documents requirements which may not be consistent with the insurance clauses.

When a contract has been let and construction work is under way, a much wider audience is involved. Senior managers and administrators may be expected to appreciate the terms used in the various parts of the documents to communicate with the appropriate language community. Section engineers, supervisers and foremen, as well as the various specialists subcontractors will have a more limited interest and will, generally, only be interested in a particular section of the Specification and of the payment documents. The terminology used and the matters implied in the various trades and among suppliers of materials will have to be related to the meanings understood by those involved.

Although the authorship and the audience may be very wide in large construction contracts, many small contracts will have a more limited audience but may well have as broad an authorship. Manufacturing or fabricating contracts or supply or maintenance contracts will be simpler, but all require, to a greater or lesser degree, a knowledge of the personnel and methods used in the industries concerned, which the draftsman must have acquired by involvement in the supervision of engineering work, and in the administration of contracts, in addition to his involvement in the design process.

On large contracts, an engineering draftsman will usually be responsible for the whole of the documentation. He will be expected to have experience of harmonising the various items of Conditions of Contract and of co-ordinating the work of measurement engineers preparing the payment documents as well as of co-ordinating the Specification with the designers involved. He will write part of the documents, probably the amendments to a standard form of Conditions of Contract, the whole of the Specification and perhaps part or the whole of the Payment Documents. In addition to his co-ordinating function, his most important duty will be to draft the special requirements and limitations and to resolve the constructional problems arising from the design in a manner which will enable a contractor to devise economical methods of construction to suit those requirements. It is all too easy for design limitations to be ignored or misinterpreted; it is the business of the engineering draftsman to ensure that those requirements are properly communicated.

1.7 The Written Utterance and The Context

The role of the written utterance is reasonably clear. In it the draftsman combines all the threads of need and intention, of technique and materials, of legal requirements and industrial rules, and of social and environmental context, to frame the basic rules required to achieve his client's purpose and bring the necessary engineering scheme to fulfilment. The contract as a whole will generally have to cover all these aspects, but instances arise where all these aspects have to be encompassed within a small part of the document writing. The following examples illustrate these points.

An example of external environment affecting only part of a document occurred when drainage had to be provided to maintain the stability of a politically and environmently sensitive hillside. This could only be carried out by wells drawing water from the superficial strata above a hard bedrock but, as the client did not wish to be burdened with long term pumping, it was necessary to provide a tunnel in the bedrock (to be constructed by a specialist contractor) to which the wells drained. Data was required concerning the effects of blasting on the hillside (so as to formulate safety limitations for the use of explosives in the tunnelling) and blasting tests had to be carried out by the contractor dealing with the hillside (as a small part of a road contract), prior to drafting the tunnelling Specification. However, these tests could not be carried out until other measures to improve stability had been undertaken by the hillside contractor and the tests had to be completed at a date which would permit the tunnelling contractor sufficient time to complete his work to meet the same completion date as the main works affecting the hillside. As a result, the document drafting involved cross references between both the Contracts and to clauses in the Conditions of Contract dealing with programme and with sectional completion. It was also necessary to provide cross references between Specification clauses describing the requirements for testing the effects of explosives and those dealing with the stability measures required before such tests could be undertaken, as well as provisions for a specialist to instrument and collect and interpret the data and for the time for completion of all the relevant operations.

In the foregoing example the external context was particularly important; political and environmental sensitivity were stressed in other clauses of both the Contract for the hillside works and the tunnelling contract, but these considerations formed only a small part of the external environment for which the documents had to be written. Account had also to be taken of the technical level of engineering methods, of the availability of materials, of current commercial considerations, of the character of available construction methods in the area concerned (particularly whether labour or machine intensive) and of many other matters which are implicit in the rules laid down in the document or in the knowledge necessary for interpretation of what has been written.

An obvious example of the direct effect of the construction methods lies in the measurements of excavation. In a context where hand excavation is the normal method, excavation in trenches will normally be in depths of 1.5 m, this being the depth that a man can satisfactorily throw material up to the next level where it is shovelled up either to the ground level or to the next 1.5 m level. On the other hand, where machine excavation is usual, excavation is measured the full depth and the maximum and minimum depths are stated. These alternatives provide rules suited to the alternative context of available construction methods.

An example of the indirect effect of social organisation on document provision arises in connection with aggregates for concrete work. The specification for such aggregate is likely to be related to the technical requirements for concrete but, in the United Kingdom which has a highly developed quarrying and sand and gravel industry it is implicit that aggregates will be provided by suppliers from existing pits or quarries and samples of the material will be available at the time of tendering. In countries such as India, where there is very little development of the aggregate supply industry, it is implicit that the aggregate will be quarried and excavated by the contractor for the work and that, for this reason, samples are unlikely to be available before the work commences.

The meaning of all documents is affected by the external context and the draftsman must be aware of such implied meanings. As an example, a requirement that communications should be by telex obviously implies a greater speed of communication than second class mail, even if it is only provided to give written confirmation to verbal instructions.

1.8 Clarity and Ambiguity

One of the most important causes of lack of clarity is the presence of ambiguities, i.e. of language capable of more than one meaning. Ambiguous words are not to be confused with homonyms, which are words with different meanings but which are spelt alike. There can be no doubt, for example, that the word "rock" used in a mechanical engineering specification refers to oscillation and not to a naturally occurring hard stratum. The context in which the word is used will generally show clearly which word is intended. There will, of course, always be some borderline cases of homonyms which are ambiguous. An example of such is the word "construction", which in an engineering contract would be expected to have the meaning "to build", rather than the lawyer's meaning "to construe". Because many of the readers of such documents may not be familiar with the latter meaning of construction, it can give rise to ambiguous interpretations; the meaning "to construe" should be avoided in engineering contracts.

Syntactic ambiguity generally arises from the incorrect construction of sentences, which gives rise to alternative possible meanings. An example is the sentence–

> The Engineer will instruct the Contractor promptly to correct the fault.

It is not clear whether the Engineer or the Contractor is to act promptly. This type of ambiguity can be overcome by careful attention to the words and phrases and the sentence construction adopted in drafting and is dealt with in Chapter 2.

Contextual ambiguity may be internal, i.e. it may arise where one provision of a document contradicts another provision in the same or a related document. Although such differences may be found by careful checking of the drafts, such ambiguities are best avoided by careful attention to the form of the document, so as to achieve a structured segregation in conjunction with appropriate cross referencing. This aspect is dealt with in Chapter 3. Contextual ambiguity may also arise from external causes. As an example, the meaning of the term "artificial obstruction" is clear within the context of a contract for dredging part of a river. Its meaning may become uncertain if a statutory authority closes part of the river not being dredged, which may temporarily prevent disposal of dredged material to the area selected by the Contractor as a spoil dump. Whether an artificial obstruction is limited to a physical item or may include a statutory obstruction will probably have to be resolved by reference to the context in which the work is carried out as well as the context of the document wording.

Clear drafting and the avoidance of ambiguities depend both on experience of drafting and of the subject matter, together with knowledge of the general character of such ambiguities and of the devices available to minimise them. Although this book explores the knowledge and the devices, it cannot, of course, be a substitute for experience.

1.9 Vagueness and Generality

Vagueness and generality are not to be confused with ambiguity. In the example of syntactic ambiguity given above, reference to acting promptly is not vague, but it is ambiguous because it is not clear whether it applies to the Engineer or the Contractor. Similarly, a reference to "hard rock" is not ambiguous, but it is vague because the degree of hardness is not defined.

Vagueness may be necessary due to the current state of knowledge concerning matters being specified or due to lack of local knowledge concerning the work to be done. This applies in civil engineering particularly to matters concerning soil mechanics, hydrology, and other branches where local knowledge may be limited, but it may also apply to items such as supervision, where advance information is insufficient or where it is inappropriate to be more precise. An example of the latter use of vagueness occurs in the phrase "superintendence shall be given by sufficient persons having adequate knowledge of the operations to be carried out"; the phraseology is deliberately vague because it is neither appropriate nor desirable in Conditions of Contract (from which this phrase is extracted) to specify the requirements more precisely. In specifications for design, an element of vagueness is essential because particular matters may not be applicable to all the possible designs for the item being specified. A common phrase in design specification "consideration shall be given to . . ." is intended to indicate a desirable characteristic to be incorporated if it can be achieved within the requirements of other criteria. Such phrases are both appropriate and essential; appropriateness and the degree of vagueness must be related to the subject matter and to the intentions which are to be communicated, matters which depend not only on circumstances but also upon the character of the drafting and the judgement of the draftsman.

Whereas ambiguity refers to uncertainty of meaning while vagueness refers to imprecision, generality deals with reference to classes of items or objects. Thus, reference to a "brass pin" is general because it neither states the type of brass nor the dimension of the item referred to as a "pin". Reference to a brass pin is neither uncertain nor vague, but allows for a variety of objects to be included within the description. Similarly, a reference to "aluminium door furniture" allows for a wide variety of knobs or handles within this general description. Generality is essential where some degree of freedom of choice or some element of variety is required and it is an essential feature in all communication. It is for the draftsman, in each instance, to adopt the degree of generality which is appropriate. In many cases overprecision is disadvantageous.

1.10 Legal and Technical Expressions

Legal expressions should be used very cautiously by engineering draftsmen. The meaning of such expressions often differs from the meaning assumed by non-lawyers; it is not uncommon for engineers to misinterpret legal expressions or to attach meanings to them which do not correspond to lawyers usage. Moreover, most of the readers and the users of engineering documents will not understand the meanings attached to legal expressions and are likely to misinterpret the intentions of clauses which incorporate such expressions. As a general rule, legal expressions should not be used other than in Conditions of Contract, and should be used with caution in that case. Engineers drafting Conditions of Contract (or amendments to them) should be fully aware of the manner in which lawyers interpret the legal expressions which are in them (not merely the definitions given in a legal dictionary) and should make a practice of adopting ordinary language whenever possible. Matters which are outside the expertise of engineers (such as clauses dealing with bankruptcy) should be the subject of legal advice.

All but a very few Latin phrases should also be avoided. Although they are much favoured by lawyers, very few engineers, managers or tradesmen concerned with engineering contracts are likely to understand them and they are unlikely to form an effective means of communication.

Technical terms are essential in all branches of engineering and construction, but they should be used only where necessary and not as a substitute for plain English. Excessive use of technical terms is likely to result in a document which degenerates into a string of technical jargon, which can confuse both the writer and the reader and lead to misunderstanding. As previously pointed out the draftsman should pay careful attention to the context in which technical terms are used; different branches of engineering and different trades make use of similar terms with different meanings. Moreover, different countries and different districts may use similar expressions with different meanings.

As an example, a joiner in the south of England is usually a tradesman who makes items such as stairs or windows in a workshop, whereas in the north of England the term is applied generally to all carpenters. As in all other matters connected with drafting, it is essential to recognise the language habits of the audience.

1.11 Interpretation in Relation to the Whole Contract

It is a legal rule that "The intention of the parties has to be collected from the whole of the contract, and particular clauses which are ambiguous must be construed in the context of the whole document". The various clauses are interdependent and should not be read separately. An apparent ambiguity could thus be resolvable in terms of the remainder of the contract. On the other hand, different parts of a contract may include contrary provisions concerning the same matter, resulting in ambiguity even though each of the clauses concerned are themselves clear when read individually. This is most likely to happen when comparing the provisions of the various documents forming the contract; unless the engineer drafting the Specification is familiar with the whole of the contents of the Conditions of Contract, he may repeat elsewhere the provisions made in the Conditions, but with a different form of words which create ambiguity.

When dealing with lengthy documents (as in many engineering contracts) it is very difficult to avoid interaction between the various clauses unless the documents are drafted in accordance with a clear and logical plan. Such a plan enables the draftsman and the reader to know where, in the document, a particular provision should be included or omitted and so enables the various provisions to be segregated in a manner which will minimise interaction. It must be recognised that a jumble of miscellaneous provisions scattered at random throughout a document is not conducive to clarity or comprehensiveness, nor is it easy to recognise, in such a document, overlapping provisions which can give rise to misunderstandings.

It must be appreciated that a logical arrangement for each document in the contract and for the relationship between the various documents is an essential means of reducing ambiguities within the whole document. The logic will vary from document to document and from contract to contract, although it is possible to formulate basic principles which will assist in arriving at logical arrangements. General matters connected with the logic and form of documents are considered in Chapter 3; subsequent chapters examine the logical arrangement and form of the various types of document which make up the usual forms of engineering contracts.

1.12 Checking

A story is told (by Max Abrahamson) of a parliamentary draftsman who, when asked by a friend what he had done that day, said, "in the morning I inserted a comma in a draft and in the afternoon I took it out again". The story is probably apocryphal, but illustrates the care which needs to be taken to produce clear and comprehensive documents. Unfortunately, engineers rarely have the time which was available to that parliamentary draftsman; in practice, documents are usually prepared to tight deadlines and within tight budgets which strictly limit the time available to the draftsman to consider the niceties which should be incorporated in the document. If the ordinary engineering firm worked in the same manner as that parliamentary draftsman it would rapidly go out of business.

All documents ought to be independently checked to identify errors and ambiguities. Because of the conditions under which the drafting of engineering documents take place there is often insufficient time for the draftsman to reread and redraft his original more than once. It is quite common for any necessary redrafting to be limited to the period needed to check the typist's transcription of the original manuscript or dictation. Because of these conditions it is essential that documents should be independently checked by a senior engineer with experience of both design and drafting, who can appreciate the requirements of the design and who can assist in eliminating any ambiguities in the text or between different parts of the document. It must be recognised that, in the course of preparing a lengthy document, the draftsman is likely to become so involved in the minutiae of design and construction as to overlook important related matters. If time is available for the document to be set aside for a period, so that the original draftsman is able to take a fresh look at it, then many of those problems can be overcome. This is, however, not usually the case and the only satisfactory solution is to provide for independent reading and examination of the documents. Estimates of cost and time in preparing the documents should make allowance for this.

CHAPTER 2

Words, Sentences and Punctuation

2.1 Avoiding Ambiguity

The use of words and phrases unambiguously is the aim of all document drafters, whether concerned with an engineering specification or a partnership deed or an Act of Parliament. The problems involved have been highlighted by various court judgements and have been analysed authoritatively by writers on legal drafting. Those writers have drawn attention to particular difficulties which have arisen and have formulated some useful rules, but have been mainly concerned with the drafting of wills, with land transactions and with legislative drafting. Nevertheless their work is equally applicable to the use of words unambiguously in engineering documents and, incidentally, to improve their readability; this chapter is concerned with translating the relevant parts of their work into terms and examples familiar to engineers.

2.2 Grammar and Usage

Many of those engaged in professional writing and document drafting have long forgotten the grammatical rules which they learned at school; they write by 'ear', based upon their current experience of the English language. Sir Ernest Gowers has pointed out that "Grammar has fallen from the high esteem that it used to enjoy" and the grammarian Jesperson has said that the grammar of a language must be deduced from a study of how good writers of it in fact write, not how grammarians say it ought to be written. George Orwell went so far as to say that "correct grammar and syntax are of no importance as long as one makes one's meaning clear". Clearly, the present view is that the writer must expect to rely upon current usage rather than the rules of formal grammar, which are based upon much earlier, and often outdated, usage.

English phraseology is not necessarily logical. Gowers points out that the common phrases 'vexed question' and 'light the fire' can be considered illogical as you cannot vex a question and a fire is something which is already alight. In an engineering context we can cite the phrases 'the second lowest tender' and "omissions made in compliance with Engineer's instructions" as illogical; logically there can be only one lowest and you cannot make an omission. All these phrases represent good current usage and there should be no hesitation in using them in documents, where appropriate. Special care must however be taken to avoid any ambiguity which might give rise to the argument that an alternative meaning is preferable because it is more logical. As an example of this, the use of a phrase such as "the second highest tidal level" has a clear meaning if it refers to a single set of tide level data; if several sets of tide level data are available, taken at different times, then the phrase might be interpreted to refer to the highest level in the second set of tidal data. An experienced draftsman would usually give the actual level rather than use a general phrase, even if this involved repeating the same level a number of times.

Although it is important to avoid using words which may themselves be ambiguous (known as semantic ambiguity), it is avoidance of ambiguity in the arrangement of words (ambiguity in syntax) which gives rise to the greatest difficulty. In the statement –

A welder may be rejected by the inspector if he has not passed the specified tests

the words may be interpreted as meaning that it is the inspector who has to pass the test, although the context will show that the tests are applied to the welder. This statement should be drafted to read –

The inspector may reject any welder who has not passed the specified tests.

Most of this chapter is concerned with avoiding syntactic ambiguities.

2.3 Choice of Words – Reference Books

The English language has a large vocabulary of synonyms and synonymous phrases from which to choose the most effective for a particular purpose. It is often difficult for the draftsman to recall from his memory the word or phrase which he needs and some assistance is then required. This is illustrated by the choice of the phrase "consistent with" in –

> The works shall be constructed in a manner *consistent with* the Employer's statutory duties.

The alternatives, such as "in accordance with" or "compatible with" were considered less satisfactory, the acceptable phrase being chosen from the alternatives given in the reference books.

Technical terms probably cause the least trouble; they tend to be precise and have few synonyms in the context in which they are used. The most useful references are the various technical glossaries available, particularly the British Standard glossaries. The Roads and Bridges Specification refers to B.S. 892 – Glossary of Highway Terms, for technical definitions applicable to road work; this is a practice which is well worth following.

The use of legal terms is often unavoidable in certain documents. It is essential, however, that engineers using legal terms should be quite sure of their meaning and the manner in which they may be used. A good legal dictionary is an essential reference, although it is no substitute for that knowledge of contract law which is today essential to the practising engineer.

A good general dictionary is an essential tool in drafting. The small pocket dictionaries are not really satisfactory and anything less than the Concise Oxford Dictionary, or its equivalent, should not be considered.

The most useful reference book for choosing the appropriate word is Roget's Thesaurus. It was first published in 1852 and consists of lists of words arranged under descriptive titles which are intended to represent the idea behind the words listed. This arrangement, together with a reference index equivalent to a large vocabulary, enables the writer to examine a whole series of comparable words and choose the most appropriate. There are today many editions based upon Roget's original arrangement; for use in drafting it is important to have a substantial edition with a large vocabulary, although this does not have to include technical terms. The smaller editions, such as the Pelican, do not give a large enough vocabulary and should be avoided.

2.4 Voice and Tense

Verbs expressing an action to be taken should be in the active voice; the passive or impersonal form may lead to ambiguity in deciding whose responsibility it is to take the action. Thus –

The Contractor shall construct diversion ways

makes it clear where the responsibility lies, whereas –

Diversion ways are to be constructed

leaves some doubt as to whether they are to be constructed by the Contractor or the Employer. This can be important if busy public roads are diverted and can give rise to substantial claims from the Contractor. Contracts are generally made in respect of obligations to be carried out in the future and should, therefore, be written generally in the future tense and in the third person. However, it is not appropriate to use complicated forms of words in order to ensure that all verbs are in the future tense; in most cases where it is appropriate to use verbs in the present or past tense it is implicit that this present or past tense is taking place in the future. It is not necessary to say –

"If the Engineer shall be of the opinion that any part of the Works shall have been substantially completed and shall have satisfactorily passed any test...". (ICE Conditions)

It is more satisfactory to say –

If the Engineer is of the opinion that any part of the Works has been substantially completed and has satisfactorily passed any final test....

Because completion must be in the future, it is necessarily implied that the action is in the future even though the verbs are in the present or in the past tense.

2.5 Arrangement of a Sentence

The variety of subject and content in engineering contracts ranges from complex legal clauses concerning the vesting of goods in the Employer, as a safeguard against the Contractor's bankruptcy, to simple instructions such as –

The Contractor shall remove from the Site, without delay, all cement which has deteriorated.

No single sentence arrangement will suit all these requirements. For the complex legal sentence, reference is often made to the rules laid down by George Coode in his essay "On Legislative Expression" published in 1843, which was intended to improve the drafting of legislation. He said that the legislative (or legal) sentence should always be arranged in the following order (the terms *case, condition,* etc. are Coode's)

(1) the *case* or circumstances with respect to which or the occasion on which the sentence is to take effect;
(2) the *condition,* what is to be done to make the sentence operative;
(3) the *legal subject,* the person enabled or commanded to act; and
(4) the *legal action,* that which the subject is enabled or commanded to do.

One of his examples is –

(1) *(case)* "Where there is any question between any parishes touching the boundaries of such parishes,
(2) *(condition)* if a majority of not less than two thirds in number and value of landlords of such parishes make application in writing,
(3) *(legal subject)* the Tythe Commision for England and Wales,
(4) *(legal action)* may deal with any dispute or question concerning such boundaries."

This form is often convenient in engineering documents as in the following example taken from a specification –

(1) *(case)* If the Statutory Undertakers or other parties having wayleaves
(2) *(condition)* desire to carry out any work on the site in connection with repairing, maintaining, taking up and relaying of gas, water and other mains, sewers, wires or cables etc.
(3) *(legal subject)* the Contractor
(4) *(legal action)* shall afford every facility for the execution of this work at all times of the day or night and in such manner as the Engineer may direct.

Many sentences can be advantageously rearranged in the form proposed by Coode. In many cases, however, either the *case* or the *condition* or the *legal subject* is stated in earlier sentences or is implied by the context. This is illustrated by the following example from a specification –

> In all cases where the cube tests required by the Clause 2.103 have not satisfied the provisions of Clause 2.104 and for other cases as may be directed by the Engineer's Representative, core specimens of 150 mm nominal diameter shall be cut from the hardened concrete of that portion of the works in question for the purpose of examination and testing.

In this case the *legal subject* (the Contractor) is implied by the context, although, because the passive voice has been used, this may not be immediately obvious. The sentence may be rearranged as follows, using the numbers to identify *case, condition,* etc.–

> (1) For the purpose of examination and testing, (2) in all cases where the cube tests required by Clause 2.103 have not satisfied the provision of Clause 2.104 and for such other cases as may be directed by the Engineer's Representative, (3) the Contractor (4) shall cut cores of 150 mm nominal diameter from the hardened concrete of that portion of the Works in question.

Coode's rules should be treated only as a guide; where an explanation of the reason for an action is included, it may be desirable to put the *condition* first, as in the following –

> In order that the softening or deterioration by exposure of the surface of an excavation may as far as possible be avoided (in excavation for foundations in clay and soft material and wheresoever the Engineer's Representative shall direct), the Contractor shall leave a bottom layer of excavation not less than 100 mm thick undisturbed, to be removed subsequently only when the layer of blinding or other concrete is about to be placed.

Many sentences, particularly in specifications, consist only of a statement or an instruction, such as –

> The site comprises the area of the Permanent Works together with the area allocated for the Contractor's workshops and stores, all as shown on the Drawings,

or –

> Demolition shall be carried out generally in accordance with the requirements and recommendations of BSCP 94 – Demolition.

The *case* and *condition* are not required in these sentences; alternatively, it may be argued that the *case* and the *condition* are implied by the context or by earlier statements. These sentences, comprising only the *legal subject* and the *legal action*, are clear and unambiguous; it is important however, to avoid inadvertently adding the *condition* at the end. This may be illustrated by the example given previously –

A welder may be rejected by the inspector if he has not passed the specified tests.

The phrase "if he has not passed the specified tests" represents the *condition* in the sentence which, when tacked on the end of the sentence, gives rise to ambiguity. An alternative rearrangement to that given previously, based upon Coode's rules would be –

(2) If a welder has not passed the specified tests, (3) the inspector (4) may reject him.

These rules are intended to assist the draftsman to develop a clear and unambiguous style. They are not intended to limit sentence formation: the one golden rule is that whatever arrangement is used in forming sentences the result should be clear and unambiguous.

2.6 Short Words and Short Sentences

Clarity of expression in English is most easily achieved by the use of short sentences and, if possible, using the ordinary relatively short words common in everyday usage. The following specification clause shows what can be achieved in this style –

All concrete surfaces shall be reasonably smooth and true and shall comply with the specified tolerances. Any fins which may occur between boards or panels shall be removed and any holes filled with 1:2 mortar as soon as possible. Any area requiring treatment, after stripping as outlined above, shall afterwards be rubbed down with a carborundum block and washed perfectly clean before the application of curing membrane. No surface treatment shall be carried out until the surfaces have been inspected by the Engineer's Representative.

Because of the need, in many cases, to qualify statements or to add provisos or exceptions, it is often difficult to achieve this simplicity. Moreover, the style is often considered inelegant because of the tradition (in legal drafting) of lengthy and complex sentences using the minimum of punctuation. The complex style is, however, gradually being ousted and the draftsman may confidently adopt the simple style of short sentences, particularly in specification writing.

2.7 Consistency of Terminology

The golden rule of drafting is – never change your language unless you wish to change your meaning and always change your language if you wish to change your meaning. For example, if reference is made to 'cost' it should not afterwards be referred to as 'price'; or if firms are described as 'hauliers' they should not, elsewhere in the document, be referred to as 'haulage contractors' unless it is intended to make a distinction between hauliers and haulage contractors. It is essential that the draftsman rigorously uses the same expression where he intends the same meaning, despite the fact that this may give rise to a monotonous style. Avoidance of ambiguity is much more important than elegant variation. If a lengthy expression has to be repeated a number of times in this way, then it is worthwhile providing a definition or a short substitute term.

When drafting a lengthy document the terminology used in an early part of the document is easily forgotten and changes in terminology can arise in a later part of the document. One of the important functions of checking the draft is to seek out such changes in terminology and to ensure that they are corrected.

This can be a tedious job when using two standard documents which were not originally drafted with the intention that they should interrelate. An example of this is the Roads and Bridges Specification which was drafted for use with the ICE Conditions and uses the terminology defined in those Conditions. If the client wishes to use GC/Works/1 with that Specification, the draftsman must either amend the definitions and the corresponding references in the Conditions of Contract or he must insert a suitable definition in the Specification such as –

Any reference to the Engineer in this Specification shall be construed as a reference to the SO.

In addition it is necessary to seek out the occasional cross reference and any use of terms different from those used in the Conditions and to individually correct them. As these differences may only amount to a handful, in a document of about two hundred pages, this is difficult to achieve without very careful checking.

The Drawings form part of an engineering contract and it is essential to ensure that terminology and instructions on the Drawings do not differ from those in the Conditions of Contract and the Specifications.

2.8 Repetition

One of the ways of ensuring consistency of terminology is to avoid repeating provisions and phraseology from earlier parts of the document. The draftsman or the typist may not repeat the provision using exactly the same words as was previously used and this will cause confusion in interpretation of requirements. The following example is taken from a specification for a nominated subcontract. It was provided that –

> If, after the Subcontractor has commenced erection, the Engineer is of the opinion that insufficient equipment, material and labour has been supplied to complete the work within the specified period, then he may instruct the Subcontractor to provide additional equipment, labour and materials and the Subcontractor shall forthwith arrange to satisfy the Engineer's requirements at his own expense.

The ICE Conditions applied to the main contract and also to the subcontract; Clause 46 of the ICE Conditions deals with the same subject and the specification clause therefore clashes with it. Moreover, it appears to give the Engineer the power to instruct the subcontractor directly instead of through the main contractor. Such repetition of contract provisions should be avoided. If there is some special reason for varying the provisions in respect of the subcontractor then this would need to be made clear by introducing the special provision with a phrase such as "notwithstanding the provision of Clause 46 of the Conditions of Contract,..." as well as stating (within the special provision) that the Engineer's instructions will be through the Contractor.

One of the commonest repetitions of phrases in specification writing is the continual restatement that work shall be carried out "to the satisfaction of the Engineer". Conditions of Contract provide that workmanship and materials shall be to the satisfaction of the Engineer and it is unnecessary to repeat this in specifications, although it must be admitted that it is sometimes difficult to avoid. If the phrase is repeated extensively it may cast doubt on the comprehensiveness of the clause in the Conditions because it can be argued that the extensive use of the phrase is intended to indicate that it does not apply where it is not used, i.e. that workmanship or materials do not have to be to the satisfaction of the Engineer where it is not specifically stated.

The practice of repetition of provisions or phrases should be avoided as it may seriously affect the interpretation of the contract.

2.9 Rules of Interpretation

In choosing his words the draftsman does not, generally, require to be guided by the rules of legal interpretation and precedent. Dickerson in "The Fundamentals of Legal Drafting" states the position simply as follows –

> "These are rules by which courts resolve inconsistencies and contradictions or supply omissions that cannot be dealt with by applying the ordinary principles of meaning. They are irrelevant because the draftsman who tries to write a healthy instrument does not and should not pay attention to the principles that the court will apply if he fails."

Lord Wilberforce has stated that –

> "The use of precedents to attribute to plain English words a meaning derived from the use of those words in other documents is always of doubtful value".

What is required is that the draftsman should choose his words and his syntax to give a clear and unambiguous meaning in ordinary plain English.

There are two questions of interpretation, however, of which the draftsman must take note.

(1) Where a list of items which may be considered to form a class is enumerated, any generalisation following those items will be limited to that class of item unless otherwise specifically stated. For example, in the phrase "concrete mixers, mortar pans, grouting pans or any other plant" the words "any other plant" will be interpreted as referring to any other plant of the type of concrete mixers, mortar pans or grouting pans; excavating plant or compacting plant will, for instance, be excluded from the description

"any other plant". This is known as the "ejusdem generis" rule. It equally affects general statements which are followed by specific requirements (see 2.10 below)

(2) Qualifying phrases apply only to the word or phrase immediately before them or following them. For example in the phrase "pipes or fittings or components of corrosion resisting material" only 'components' would be interpreted as being of corrosion resisting material. This is discussed in 2.11 below.

2.10 Elaborating a General Statement

When a general requirement is included in a document it is often desirable to emphasise particular aspects of that requirement. Thus, on the subject of noise control –

> The Contractor shall comply with the general recommendations set out in BS 5228 – Code of Practice for Noise Control on Construction and Demolition Sites. All vehicles and mechanical plant used for the purpose of the Works shall be fitted with effective exhaust silencers and all compressors shall be sound reduced models fitted with properly lined acoustic covers.

The ejusdem generis rule referred to in 2.9 (1) applies in this case also, i.e. the interpretation of this clause is likely to be that the requirements to comply with BS 5228 are limited to the sound reduction of construction plant, whereas the draftsman intended that the whole of the requirements of BS 5228 should apply and also that particular measures should be taken for the sound insulation of vehicles, mechanical plant and compressors. The solution is to say that the general statement is not to be limited by the particular requirements; the second sentence should be redrafted as follows –

> *Without prejudice to the generality of the foregoing*, all vehicles and mechanical plant used for the purpose of the Works shall be fitted with effective exhaust silencers and all compressors shall be sound reduced models fitted with properly lined acoustic covers.

Similar considerations apply where limitations are to be inserted which might be interpreted as affecting other provisions of the contract. This situation might arise in connection with the noise control clause referred to above if limitations are imposed on working hours of plant. Most contracts have provisions concerning general working hours (both daily and weekly) and, to ensure that these are not affected by limitations applied solely to plant, the form of drafting should be –

Without in any way limiting any other applicable provisions of the Contract, the working hours of plant shall be restricted to Mondays to Fridays between 07.30 hours and 19.00 hours and Saturdays between 07.30 hours and 13.00 hours.

The phrase referred to in 2.9 (1) above can be similarly dealt with if it is intended to preserve the generality of "any other plant". Works on legal drafting suggest that one should add the following sentence –

The generality of the words 'any other plant' is not to be limited by the preceding particular words

This phraseology is rarely used in engineering documents. The more usual practice is either to provide particulars covering sufficient classes of plant, thus avoiding any practical limitation of the phrase "any other plant", or to redraft with a linking phrase removing the limitation, thus –

"Any plant *including but not limited to* concrete mixers, mortar pans, grouting pans, and the like."

The phrases which have been italicised represent the standard practice for avoiding limitations of generality resulting from the ejusdem generis rule.

2.11 Adjectives Qualifying Two or More Nouns

The rule in 2.9 (2) applies to all cases in which an adjective qualifies two or more nouns, although the resultant ambiguity may be resolved by the context. For example, the phrase –

Illuminated signs or bollards

taken without any textual references might be interpreted as referring to illuminated signs and to bollards rather than to illuminated signs and illuminated bollards and might give rise to claim for additional payment if a subsequent direction is given to provide illuminated bollards. If the phrase in context reads –

Illuminated signs or bollards, required to provide visibility at night then it can be interpreted as a reference to illuminated signs and illuminated bollards. However, it is incumbent on the draftsman to provide phraseology which is clearly unambiguous; redrafting as –

Illuminated signs or illuminated bollards

or as –

Signs or bollards (both illuminated)

will rectify this.

A similar connecting phrase can be used in the example given in 2.9 (2), i.e. it can be redrafted as –

Pipes or fittings or components, all of corrosion resisting material,

so ensuring that the phraseology is not misinterpreted.

The interpretation of more complex sentences may baffle even the best legal minds, as witness the following from an early Motor Car Act of Victoria, Australia –

> "Every person who drives a motor car on a public highway recklessly or negligently or at a speed or in a manner which is dangerous to the public *having regard to all the circumstances of the case including the nature condition and use of the highway and to the amount of traffic which actually is at the time or might reasonably be expected to be on the highway* shall be guilty of an offence against this Act."

In one case the judge held that the words which are italicised modify "recklessly", "negligently", "at a speed" and "in a manner"; in another case the judge held that the phrase did not apply to "recklessly". A similar argument could arise in connection with the following contract requirement –

> The Contractor shall make it a condition of employment that his employees shall not park cars adjacent to the Site nor congregate in, nor play games in the streets adjoining the Site in a manner which will cause a nuisance to local residents during Site working hours.

Although the phrase "during Site working hours" clearly applies to playing games in the streets adjoining the site and probably to congregating, does it apply to parking cars adjacent to the Site? If it does not, then employees are not permitted to park adjacent to the Site, even outside working hours. If this is not the intention then the sentence should be redrafted as follows –

The Contractor shall make it a condition of employment that employees shall not:
   (a) park their cars adjacent to the Site; nor
   (b) congregate in nor play games in the streets adjoining the Site;
        in a manner which will cause a nuisance to local residents during Site working hours.

2.12 Use of Pronouns

Pronouns are very useful to the draftsman as a means of improving readability; continual repetition is wearying to the eye and may, in practice, give rise to misunderstandings. However, the prime requirement is to avoid ambiguity; the noun should, therefore, be repeated if the use of a pronoun may result in ambiguity.

There are two situations in which pronouns should particularly be avoided:—

(1) Where the pronoun becomes divorced from the reference phrase, as in the following –
   Each cofferdam in the river shall be provided with a sluice gate of appropriate size conveniently situated, together with operating headgear and platforms, as necessary to enable it to be flooded in case of emergency.
   The pronoun "it" is intended to substitute for cofferdam not for sluice gate (flooding the sluice gate would be a very odd requirement) and it is better to use the noun "cofferdam" in this instance rather than use the pronoun.

(2) Where two or more nouns occur in the same sentence ambiguity may arise due to a doubt as to which noun the pronoun applies. This is a very common problem when such names as "the Engineer" and "the Contractor" are referred to as in –
   The Contractor shall make provision for the Engineer to inspect the source of granular fill if he so desires.
   Does the pronoun "he" apply to the Contractor or to the Engineer, i.e. is the Contractor entitled to conclude from the phraseology that he only needs to make provision for inspection if he desires the Engineer to inspect the sources of granular fill? The draftsman intended the desire to emanate from the Engineer and

he should therefore have reused the noun "Engineer" in order to avoid ambiguity.

2.13 Singular and Plural, Masculine and Feminine

Questions sometimes arise as to which pronoun to use when referring to a group. Thus, the Engineer may be a partnership or (as often occurs in international contracts) may consist of two parties – one responsible for technical matters and the other for financial matters. Today, in all but the smallest contracts, the Contractor is likely to be a limited company or corporation, or may be a partnership of limited companies or a consortium. Similar considerations apply to Employers and to public authorities. Furthermore, the gender and the number of persons forming a party or entity referred to in the contract, may change during the course of the contract.

Although there is no rule concerning the pronoun to be used the convention normally adopted is:—

(a) The number of the pronoun follows the number of the noun it replaces, i.e. the Engineer or the Contractor (and other similar entities) are considered singular persons, but 'the Harbour Commissioners' and other similar entities are considered plural.

(b) All human entities are considered masculine. There is no reason why the alternative of considering them feminine should not be adopted as long as the same convention applies throughout the contract documents so that any change of gender during the period of the contract will not affect the interpretation of the contract.

(c) The pronouns 'he' and 'they' are used only to describe persons; inanimate objects are referred to as 'it', even though such objects (as, for example, a ship) are commonly referred to as he or she.

Section 61 of the Law of Property Act 1925 confirms a number of conventions. It provides that –

"In all deeds, contracts, wills, orders and other instruments executed, made or coming into operation after the commencement of this Act, unless the context otherwise requires –
(a) "Month" means calendar month;
(b) "Person" includes a corporation;

(c) The singular includes the plural and vice versa;

(d) The masculine includes the feminine and vice versa."

Despite the conventions, it is unreasonable in a modern context to use the masculine or feminine gender where it is clearly not required. The most obvious example of this is in the Form of Tender attached to the ICE Conditions which commences with the words –

"GENTLEMEN, Having examined the Drawings, Conditions of Contract,..."

The word GENTLEMEN can be omitted without in any way affecting the sense of the document; its use becomes ludicrous on government contracts addressed to the Secretary of State when that person is a woman.

2.14 The Use of Capital Letters

In ordinary English usage capital letters are reserved for the first letter in a sentence and for the first letter of a proper noun. In legal documents defined words and phrases are treated as proper nouns; throughout the document those words in every case have capital letters, as in the phrase "Date of Commencement of Works" which is defined in the ICE Conditions. It is, of course, essential to confine the capital letters to those defined words or phrases and to avoid using capital letters for the same words or phrases used in the non-defined sense, as italicised in the following sentence –

The Contractor shall co-operate and co-ordinate with the *contractor* constructing the adjoining section of the road...

Generally the definite article does not have capitals; where it forms part of the definition it must be used wherever the defined term is used.

2.15 "And" and "Or"

Legal draftsmen recognise that "and" and "or" have many shades of meaning and that their careless use may give rise to ambiguities. Many of these difficulties arise in wills and similar documents which may have lengthy complex sentences involving relationships between many living persons and their offspring. Short simple sentences reduce the risk of such ambiguities; they are, therefore, a lesser problem in engineering documents. Nevertheless, even in simple sentences it is very easy for ambiguities to arise which may result in misinterpretation and consequent claims for extra costs. Many of the apparent ambiguities can be resolved by reference to the context in which the statements are made, but such reference should be treated cautiously; it is preferable that the words and phrases used be unambiguous rather than that it be necessary to demonstrate the meaning by reference to context.

The use of "and" and "or" is so extensive that it is unlikely that all possible ambiguities can be catalogued; the following paragraphs set out the main problem areas and pitfalls.

In ordinary use "and" has a conjunctive effect, e.g. a reference to "roads and streets" in a specification describes a class which is the subject of that specification. In some contexts it can have both a conjunctive and a disjunctive effect as in the phrase "The Contractor shall...provide... transport to and from and in and about the Site..." which appears in the ICE Conditions. Does the phrase "transport to and from" mean that such transport is limited to journeys which are both to and from the Site or does it mean "journeys to the Site or journeys from the Site or journeys both to and from the Site"? It can be deduced from the context that the latter statement is intended; it would have been better to refer to "to or from and in or about the Site", since providing for transport to and transport from must necessarily include transport both to and from. Although, due to the context, the phrase is unlikely to give rise to a claim in this instance, there are occasions where the use of "to and from" (instead of "to or from") could give rise to a claim that 'transport to' alone and 'transport from' alone is not included in the contract and should be dealt with separately, particularly where hired transport is used for single journeys involving extra costs for returning to the hire yard.

Similar possibilities for ambiguity arise where "and" is used between two or more adjectives. In a contract for site supervision reference is made to the managing contractor providing "civil and structural engineers". Does this mean the provision of civil engineers and structural engineers or of engineers who have both civil and structural qualifications? Once again context is important but, in this instance, if two different types of engineers are required it would be preferable to say "civil engineers and structural engineers". A more complex problem arises where three adjectives are involved as, for example, where a design and construct invitation to tender requires the tenderers to offer "simple, labour intensive and economical designs". Does this mean that acceptable designs may be –

> simple but not labour intensive or economical, or labour intensive but not simple or economical, or economical but not simple or labour intensive, or simple and labour intensive but not economical, or labour intensive and economical but not simple

or is it intended that "designs shall be economical and also both simple and labour intensive".

It should be noted that "and" in a positive statement changes to "or" in the negative. Thus, "the ground contains both sulphates and chlorides", but, "the ground does not contain either sulphates or chlorides".

The use of "or" is usually intended to be disjunctive but a conjunctive meaning may be implied as, for example, in the phrase "transport to *or* from the Site" which necessarily implies also "transport to *and* from the Site". The disjunctive meaning can be emphasised by the use of "either...or", but this applies only in the case of two items; for three or more it is necessary to use the phraseology "or alternatively". Thus, we refer to –

> "either A or B", in the case of two but "A or B or alternatively C..." in the case of three.

Even with this terminology it is necessary to appreciate what is being excluded. The phrase –

> drainage pipes shall be either clay or concrete,

does not mean that all pipes shall be either clay or concrete, merely that a pipe shall not be a combination of the two. It is, therefore, within the meaning of the phraseology if some of the pipes are of concrete and some of clay. If it is intended that different types of pipe shall not be mixed then it should be stated as –

> all drainage pipes shall be either clay or concrete

If, on the other hand, there is merely objection to a mixture of pipes it may be necessary to put a limitation on this by adding that (as in the Roads and Bridges Specification) –

> Only one type of pipe shall be used within any drain length between manholes.

In some cases "and" can be substituted for "or", with a small difference in the shade of meaning. As an example, a specification provided that –

The instruments shall be calibrated at the beginning and completion of each day's survey *or* more frequently if required by the Engineer.

This suggests that, if a more frequent calibration is ordered, then calibration at the beginning and completion can be overridden, i.e. a situation could result in which calibration was carried out four times in a day but not including calibration at the beginning and end of each day's working. Alternatively, if *and* is substituted for *or* then this suggests that the increased number of calibration checks must include checking at the beginning and at the completion of each day's survey.

Occasionally it is advantageous to use "and/or", although, due to its past misuse, it is often deprecated. In the specification provision –

Soil for backfilling into trenches shall be granular material or firm clay from the excavation...

the intention is that either or both materials could be used; so the requirements could, with advantage, be rewritten –

Soil for backfilling into trenches shall be granular and/or firm clay....

Such usage should be treated with caution; it has been found that, in many cases, the use of "and/or" can be misleading, especially when more than one of these terms appears in a sentence. The worst culprits for misuse of "and/or" have, in the past, been writers of insurance policies. Piesse (in "The Elements of Drafting") gives the following example of the horrific use of this term in a marine policy referring to goods to be shipped –

"at and from port or ports, place and/or places in the United Kingdom and/or continent of Europe and/or United States of America and/or Canada to port or ports place and/or places in Australia and/or Tasmania."

He also gives an example of a nonsense created by the use of "and/or" as follows –

"If the ship is totally destroyed by fire in the Pacific Ocean and/or Mediterranean"

This is obvious nonsense when read with "and". Although it is unlikely that such phraseology will be used in documentation of a modern engineering contract, it is worth taking note of the consequences of excessive use of "and/or".

An alternative to "and/or" is the addition of the words "or both" as in –

> Soil for backfilling into trenches shall be granular material or firm clay from the excavation or both...

Other more esoteric uses of "and" and "or" have been considered by the courts, involving complex phraseology of multiple adjectives and nouns, but these are of little interest to engineer draftsmen.

2.16 "Shall" and "Will"

It has been common practice among legal draftsmen to use the word "shall" not only as an imperative to state what shall or shall not be done but also to define circumstances, such as in the phrase "If the Engineer shall be of the opinion...". As long ago as 1843 George Coode (in his essay referred to in 2.5 above) deprecated this usage; the same objections are raised by modern writers on legal drafting. A résumé of the arguments is given in 2.4 above under the heading of Voice and Tense. Draftsmen of engineering documents are strongly advised to use "shall" only to state what shall or shall not be done and by whom and always to use it (or to use the word "will") in that context.

Engineering contract documents generally serve two purposes:

(a) to be issued to contractors as the basis of a tender for performing the work detailed in the document; and/or

(b) as part of a contract for the construction of the Works.

When used for tender purposes, the documents must contain statements of the information and physical provisions to be made available or to be given to the Contractor by the Employer or the Engineer. Such items might read –

> Details of the lands, owners and tenants affected by the Works will be provided by the Employer

or –

> The survey monuments have been constructed in the positions shown on the Drawings and will be made available by the Engineer...

The use of the word "will" in the document, for such actions by the

Employer or the Engineer, is appropriate and is usually carried over into the Contract without alteration. This practice has, therefore, become commonplace and is often adopted even when no tendering is involved.

The comments on the use of "shall" are not intended to exclude the use of "may" when the action referred to is not obligatory.

2.17 "Such", "Said", "Same", "Because of"

These are all words which should either be avoided or be used to only a very limited extent. In many cases they merely represent ancient forms of legalese which are gradually going out of use.

The construction "such...as" used, for example, in the following extract from Clause 7 of the ICE Conditions –

> "The Engineer shall have full power and authority to supply and shall supply to the Contractor from time to time during the progress of the Works *such* modified or further drawings and instructions *as* shall in the Engineer's opinion be necessary..."

is a proper usage of English and is not included in the strictures on the use of "such". However its use as a demonstrative in phrases such as "such person" or "such material" should be limited to where it is essential as a means of discrimination between a defined noun and the general use of that noun. The following clause from a specification is an example of inappropriate use –

> The Contractor shall give to the Engineer written notice of the preparation or manufacture at a place not on the Site of any pre-constructed units or parts of units to be used in the Works, stating the place and time of preparation or manufacture so that the Engineer may make inspection at all stages of the work. *Such* notice shall be given at least two weeks before preparation or manufacture is due to commence.

In this case "such" is merely a substitute for "the"; the clause would be better drafted as follows –

> The Contractor shall give to the Engineer written notice at least two weeks before commencement of preparation or manufacture...

The second sentence of the clause could then be omitted. An appropriate use of "such" as demonstrative occurs in Clause 17 of the ICE Conditions which says –

> "If at any time during the progress of the Works any error shall appear or arise in the position levels dimensions or alignment of any part of the Works the Contractor on being required so to do by the Engineer shall at his own cost rectify *such* error to the satisfaction of the Engineer unless *such* error is based on incorrect data supplied in writing by the Engineer's Representative in which case the cost of rectifying the same shall be borne by the Employer".

In this case "such" distinguishes between error in "the position levels dimensions or alignment" as distinct from other errors which may occur. In general, it is preferable to avoid the use of "such" in these contexts unless its use is seen as essential discrimination; it can often be avoided by the use of a definition.

The use of "said" as a demonstrative has, except in legal terminology, disappeared from English usage; the authorities on legal drafting deprecate its use and recommend that it be avoided, as it is often a mere superfluity or can be replaced by a pronoun. For example "the *said* material" can be replaced by "it" without affecting the meaning of the sentence. Draftsmen of contract documents should avoid any use of the word or of its derivative "aforesaid".

The same remarks are applicable to "the same" used in a demonstrative sense. In the sentence from Clause 17 of the ICE Conditions, referred to above, the phrase –"the cost of rectifying *the same* shall be borne by the Employer." can be rewritten as –"the cost of rectifying it shall be borne by the Employer."

"Because of" is not a phrase often used in contract documents and is best avoided because of the risk of ambiguity in certain contexts. The requirement that –

> The Contractor shall not remove protective barriers because of obstructions to personnel movements

is an example of such ambiguity. It is not clear whether the draftsman means that the Contractor shall not remove protective barriers even if such barriers cause obstruction to personnel movements or whether the removal of the protective barriers will result in obstruction to personnel movements. In the former case it would be preferable to write –

> Obstruction to personnel movements shall not constitute a reason for the Contractor to remove protective barriers.

In the latter case it would be better to write –

> The Contractor shall not remove protective barriers so as to cause obstruction to personnel movements.

2.18 Special Phrases

Examples have already been given of some special phrases such as "including but not limited to" which are useful as connecting phrases. The following phrases frequently occur and should form part of the draftsman's vocabulary –

(a) "Shall comply with" as in – "The Contractor *shall comply with* the regulations and requirements of the Electricity Board".

(b) A similar phrase which is used in referring to matter included in the document or in another document (rather than in regulations which require compliance) is – "in accordance with the provisions of Clause...". Where there is a possibility that some of the provisions may not be applicable, it is usual to say "in accordance with the relevant provisions of..."; this commonly applies to British Standards.

(c) If it is intended to make an exception to some earlier requirement in the document, then the phrase to be used is "notwithstanding the provisions of Clause...". The use of this phrase is important to avoid conflict between different provisions of the contract, particularly between a requirement in the Specification (which is specific) and a requirement in the Conditions (which is general).

(d) Where it is intended that a matter referred to shall be conditional, the statement may be preceded by the words "subject to" as in "The Employer shall (*subject to* this clause) serve all such notices...".

(e) Where there is to be a choice from a list the phrase used should be "any of the following–"; where the list is exclusive it may be followed by the phrase –"but no other", as in the provision that – "corrosion protection coatings shall be obtained from *any of the following* manufacturers *but no other*".

(f) Where an operation is not continuous the phrase–"from time to time" may be used, as in "The Contractor shall from time to time provide such assistance as the Engineer may prescribe". It can also be used, in certain circumstances, to discriminate between alternatives, as in the provision that "The Contractor shall *from time to time* maintain, reconstruct or divert rights of way...".

(g) Another phrase for discriminating between alternatives is "in either event" as in "If the approach is obstructed or the lock gates damaged then *in either event* the Contractor shall...".
If there are more than two events involved then the phrase becomes "in any such event".

(h) Where there is a provision that may be interpreted as affecting other important matters for which the party concerned is responsible, then the usual phrase is "this requirement shall not in any way relieve.......of his duties or responsibilities under the Contract".

2.19 Phrases Concerning Time

Many engineers are not aware that if a period of time is specified "from" a particular date then that date is not included in the calculation of the end date of the period, i.e. seven days from first January will be eighth January. The ICE Conditions state that the Time for Completion shall be calculated from the Date for Commencement of the Works, i.e. if the Works are to be completed within three hundred days then the Contractor has three hundred and one days to carry out the work. If it is intended to include the initial date within the period then the phrase should be "from and including", or alternatively "on and from" or "on and after". In modern statutes the phrases "commencing with" and "ending with" a named day are used; these expressions leave no doubt that the day referred to is included within the period.

If a period is to commence "after" a particular date then that date is not included; this applies also to the expression "within" a specified period, after a particular date.

Use of the word "until" a particular date is ambiguous as also is the word "to". If the date is to be included the phrase "until and including" should be used; alternatively "from date A to date B, both days included".

If work is to be completed "by" a specified date then the day having the date is included.

The meaning of day, week and year are clear, but month (by law meaning a calendar month) can give rise to difficulties, and is best avoided, as it varies between twenty eight and thirty one days. Contract periods are best stated in weeks or, if they are short, in days.

When stating periods before and after a particular date care must be taken to avoid the "undistributed middle". For example, if part of the Works is to be completed before a certain date and the remainder to be completed after it then that date is not included in either period.

Disagreements often occur when words such as "forthwith" or "immediately" are used. It is preferable to use a specific period if possible, as for example "within twenty four hours" or perhaps "within four hours" if it is a matter of greater urgency. However, it is difficult to generalise on this point and in many cases it is still necessary to use a general term, the length of time then depending upon the circumstances of the case.

2.20 Punctuation

The older legal writers drafted documents without any punctuation, even omitting full stops and writing from edge to edge of a parchment. This practice has been gradually modified, but even in the nineteenth century it was common to omit all punctuation except full stops and occasional parentheses. The courts tended to ignore punctuation in construing documents, based upon a judgement in 1816 in which the Master of the Rolls stated –

> "It is from the words, and from the context, not from the punctuation, that the sense must be collected".

This view no longer prevails. Lord Shore in a House of Lords judgement in 1918 stated –

> "Punctuation is a rational part of English composition, and is sometimes quite significantly employed. I see no reason for depriving legal documents of such significance as attaches to punctuation in other writings."

The attitude is summed up in the following quotation from Burrows "Words and Phrases Judicially Defined" –

> "The principle applied appears that the courts will use punctuation

marks appearing in the document and, whenever necessary, supply them, but such marks are only used to assist in understanding the words. If they are without sense or conflict with the plain meaning of the words in the document, the courts will not allow them to cause the meaning to be placed upon such words which they otherwise would not have."

Modern Acts of Parliament are fully punctuated, as are most modern documents. Conditions of Contract are generally fully punctuated although the ICE Conditions have been drafted without any commas, except where these appeared in a quotation from the Fair Wages Resolution of the House of Commons. The modern tendency is to limit punctuation marks to essential positions and avoid undue proliferation.

In speech we can vary the meaning of the words by the tone and level of the voice and by the position and length of pauses. To a limited extent, punctuation performs a similar function in writing. The actual punctuation used depends on the writer and on the style and purpose of the writing. For example, in this book where a quotation or example is inserted in or at the end of a sentence to illustrate a point, then it is preceded by a dash; if the quotation or example is lengthy, then it starts on a new line and is inset. This is not to be confused with punctuation adopted for marshalling and for listing as explained in Chapter 3. The reader will find that there is considerable variation between the punctuation adopted by different writers and that literary works (which do not use marshalled sentences and rarely have lists) will adopt a different style of punctuation to that used in documents.

There is little to be said about the use of full stops to indicate the end of a sentence except to note that its use with isolated sentences in headings and the like is becoming uncommon and should be avoided. Question and exclamation marks are never used in document drafting and do not, therefore, require comment here.

2.21 Colon, Semicolon, and Dash

The colon and the semicolon were, in the past, considered almost interchangeable and having a 'strength' between a comma and a full stop, the colon having a little more 'strength'. In modern practice the functions of the colon have been largely taken over by the semicolon (particularly in document drafting), the colon being restricted to specialist functions.

In drafting practice the use of the colon is usually limited to:

(a) indicating the commencement of item enumeration (or marshalling) within a sentence, as for example –
> ...the following provision shall take effect:
> (i) all sums of money that may be due...;

(b) a mathematical symbol for 'is to', as for example –
> ...a mixture of pulverised fuel ash and Portland Cement in the proportions 15:1...;

(c) a slightly heavier stop than a semicolon, equivalent to the use of the word namely, for example –
> the requirements for ducts will vary with depth: at a depth less than four feet and greater than ten feet they shall be surrounded with concrete Class 22.5/20.
> This use is rather rare; an alternative phraseology will avoid it.

The semicolon is most useful for connecting two related requirements, where framing them in consecutive sentences might give rise to the impression that they might not be connected. It is common to use the conjunction *and* for this purpose, but the alternative use of a semicolon will often obviate ambiguities which may arise through the use of *and*. As an example, the following sentence appears in GC/Works/1 –

> "Upon the completion of the Works to the satisfaction of the SO the Contractor shall be entitled to be paid the amount which the Authority estimates will represent the Final Sum less one half the amount of the reserve, and thereafter the Authority may, if he thinks fit, pay further sums in reduction of the reserve"

As an alternative a semicolon has advantages in place of *and* thus –

> ...the Final Sum less one half the amount of the reserve; thereafter the Authority may, if he thinks fit, pay further sums in the reduction of the reserve.

The following example from a specification illustrates the use of more than one semicolon to connect several common requirements into one sentence –

> References in LN 139 to concrete quality A shall be deleted and concrete Class 22.5/20 substituted; references to concrete qualities B and C shall be deleted and concrete Class E substituted; references

to concrete quality F shall be deleted and concrete Class 30/10 substituted.

The dash is most commonly used at the commencement of item enumeration (or marshalling), either with a colon or on its own. Which alternative the draftsman chooses to adopt is a matter of taste; a comparison of the following quotations illustrates this.
From ICE Conditions –

"For the purpose of this Clause:—
(a) the expression "the Act" shall mean..."

From GC/Works/1 –

"In this Condition –
(a) the expression 'loss of property' includes...".

For a discussion on drafting in the form of item enumeration or "marshalling", and also of "listing" the reader is referred to Chapter 3.

Quotations and examples are rarely inserted in sentences in contract documents, although they may form part of the data provided. The use of the dash for this purpose (referred to in Part 2.20) is not, therefore, relevant to document writing.

2.22 Commas, Brackets, Inverted Commas, and Hyphens

Next to the full stop, commas are the most useful punctuation mark but also the most subject to misuse. They should not be regarded as a substitute for an unambiguous arrangement of the wording of a sentence, but rather as a device to assist readability, particularly in complex sentences. The ICE Conditions are drafted without a single comma, but the FIDIC Conditions (which originally had almost identical wording) have been punctuated with commas; the readability of the FIDIC Conditions improved as a result.

The most obvious use of commas is to separate items listed in a sentence. Thus –

Unless otherwise specified, all lighting units shall use fluorescent lighting and be complete with switch gear, starters, ballasts, fuses, and capacitors.

obviously requires commas between the various items to clearly distinguish between them. A different arrangement of commas or no commas might alter the meaning in the following way –

...with switch, gear starters, ballasts, fuses and capacitors.

In this alternative, the comma before *and* has been omitted and it is no longer obvious whether fuses are separate from capacitors or form a combined unit. It has recently become a fashionable practice not to use a comma before *and*. Authorities on punctuation and legal drafting do not support this view; it is considered that where there is uncertainty as to whether the last two items in a list are combined or separate then a comma should be used before the final *and*. Carey (in "Mind the Stop") quotes the following instance where the use of a comma before *and* was essential –

> "Five London boroughs...Barnes, Hammersmith, Brentford and Chiswick, Fulham, and Wandsworth."

Without the final comma the reference could be to 'Fulham and Wandsworth' as a single borough similar to 'Brentford and Chiswick'.

Although a comma is obviously necessary between consecutive items in a list, no comma is needed between a number and the following word. It was previously common to place a comma between a number and a street name in an address but it is now acceptable to write, for example, 59 Southwark Street in a formal address or in an address given in a document.

Where a descriptive phrase is inserted in a sentence it is appropriate to insert a comma before and after the phrase, to differentiate it from the phrase being described. Such descriptive phrases are generally not essential to the sentence formation; they may be omitted from the sentence and made the subject of (or part of) another sentence, but in practice this usually proves cumbersome. Alternatively, the descriptive phrase may be placed within brackets. The following sentence illustrates the point –

> Any drainage sumps required shall, where practicable, be sited outside the area excavated for the drainage Permanent Works...

The "where practicable" can be made the subject of a separate sentence thus –

> Any drainage sumps required shall be sited outside the area excavated for the drainage Permanent Works. Where this is not practicable the Contractor will be permitted to site the sumps within the area excavated for the Permanent Works.

This additional sentence is obviously not justified; the original sentence deals with the matter equally well and more concisely, thus improving readability. Alternatively, the phrase may be enclosed in brackets, as follows –

Any drainage sumps required shall (where practicable) be sited outside the area excavated for the drainage Permanent Works.

Generally, brackets are used where the descriptive phrase is less directly connected to the phrase being described, but this is largely a matter of judgement and, in many cases, either alternative is appropriate. An example (from the same specification) where brackets were used illustrates this –

Pipes of nominal diameter greater than 300 mm may (at the Contractor's discretion, unless otherwise provided in the Contract) be clay or concrete.

The matter is succinctly summed up by the definition of 'parenthesis' in the Concise Oxford Dictionary – "Word, clause, or sentence, inserted into a passage to which it is not grammatically essential, and usually marked off by brackets, dashes, or commas". Dashes are rarely used in documents for this purpose and this use may be ignored. Bracketed sentences are occasionally used in the form of explanatory notes or as an explanatory description in a title, but which does not form part of that title.

A further use of the comma is to separate phrases which, if not separated, would give rise to some confusion in the mind of the reader. This may be illustrated from a well known standard specification which contains the sentence –

When required, concrete for post footings shall be Class E concrete to Clause 1602.

If the comma after "required" is omitted the sentence could be read as follows (commas have been inserted for illustrative purposes only) –

When required concrete, for post footings, shall be Class E concrete to Clause 1602.

This can give rise to temporary confusion in the mind of the reader; the comma after "required" is, therefore, needed to assist readability. The misuse of commas for this purpose can give rise to even greater confusion; such commas are usually inserted 'by ear' and caution is necessary to avoid excessive use. The following sentence gives a good illustration of the various uses of commas and brackets and of the balance to be maintained in their use –

> The notice of suspension shall specify the event or events concerned and the date from which the suspension shall be deemed to take effect (which effective date may be earlier than the date of the notice by not more than 28 days) and the suspension shall be effective until such date as the Engineer shall notify the Contractor that the temporary suspension of the formula is cancelled, which date of cancellation shall be when, in the opinion of the Engineer, the effect of the event or the events ceases.

Inverted commas are used to indicate words, phrases or sentences which are quoted and should be used only at the beginning and end of the actual words quoted; it is not permissible to insert other words than those which occur in the original quotations. If some words are omitted, then this is indicated by dots thus ...; these omissions may occur at the beginning, the middle or the end of the quotation. Where an omission occurs in the middle of a quotation, the draftsman must be careful to avoid distorting the original meaning; such a distortion could mislead one of the parties to the contract. There appears to be no definite rule concerning the use of double inverted commas or single inverted commas, but double inverted commas are commonly used at commencement and end of a quotation, while single inverted commas are used for a quotation within a quotation. Inverted commas are often used where a title or a defined term is quoted although, in this case, it is more common to use capital letters in document drafting (as explained in 2.14 above). An important use is to distinguish ordinary words which are used as names or quoted descriptions; for this purpose single inverted commas are more common in document drafting, as in the following sentence –

> Vitrified clay pipes for foul sewers and surface water drains shall comply with the relevant requirements of BS 65: 1981 and be of 'normal' type with flexible mechanical joints.

A similar but less common use is when quoting standard phrases such as the phrase 'by ear' used above.

The sole use of hyphens (which must not be confused with dashes) is to form compound words such as 'push-fit' or 're-made' or 'de-watering'. When such compound words become commonplace the hyphen is dropped, as in 'bypass'. At an intermediate stage, either use may be acceptable thus 'dewatering' is sometimes used in place of 'de-watering'.

2.23 Provisos, Exceptions, Conditions, and Qualifications

The phrases "Provided that" or "Provided always that" occur only in legal documents; the provisions which follow these phrases are referred to as 'provisos'. They are intended to be applicable where a general rule needs to be varied or modified in its application to particular circumstances, but are often liberally sprinkled through a document to introduce exceptions, qualifications of statements and various applicable conditions. Coode describes the proviso as "that bane of all correct composition". All the modern authorities on legal drafting recommend their avoidance, particularly as this does not involve any great difficulty. Two examples of provisos used in Conditions of Contract illustrate this. In GC/Works/1 a lengthy clause dealing with 'determination' has a proviso inserted at the end as follows (the central part of the clause, which is not relevant to the example, has been omitted) –

"The Authority may without prejudice to the provisions contained in Condition 46 and without prejudice to his rights against the Contractor in respect of any delay or inferior workmanship or otherwise, or to any claim for damage in respect of any breaches of the Contract and whether the date for completion has or has not elapsed, by notice absolutely determine the Contract in any of the following cases, additional to those mentioned in Condition 55 hereof;.........................is prejudicial to the interest of the State: Provided that such determination shall not prejudice or affect any right of action or remedy which shall have accrued or shall accrue thereafter to the Authority."

In this case the words 'provided that' appear to be superfluous; there appears to be no good reason why the end of the clause should not be written –

...is prejudicial to the interests of the State. Such determination shall not prejudice or affect any right...

The rather peculiar punctuation used in GC/Works/1 (colon followed by a capital letter) does not appear in the next example, taken from the ICE Conditions –

"In the event that the Nominated Sub-contractor shall be in breach of the sub-contract which breach causes the Contractor to be in breach of contract the Employer shall not enforce any award of any arbitrator or judgement which he may obtain against the Contractor in respect of such breach of contract except to the extent that the Contractor may have been able to recover the amount thereof from the Sub-contractor. Provided always that if the Contractor shall not comply with Clause 59B(6) the Employer may enforce any such award or judgement in full."

There is some justification for this use of the proviso, which can be described as varying the general rule in respect of particular circumstances. Nevertheless, it is reasonably easy to redraft it without the proviso, as follows –

In the event that the Nominated Sub-contractor is in breach of the sub-contract, which breach causes the Contractor to be in breach of contract, the Employer shall not enforce any award of any arbitrator or judgement which he may obtain against the Contractor in respect of such breach of contract, except to the extent that the Contractor may have been able to recover the amount thereof from the Sub-contractor, unless the Contractor fails to comply with the Clause 59B(6) in which case the Employer may enforce any such award or judgement in full.

The redrafted clause has been punctuated with commas for ease of reading.

Reed Dickerson in his book 'The Fundamentals of Legal Drafting' says –

"Provisos have been used for so many purposes (to state conditions or exceptions or simply to add material) that they tend to be ambiguous. At best they constitute archaic legalisms. Accordingly, provisos should be avoided."

This advice is equally applicable to draftsmen of engineering documents.

To avoid the risk of ambiguity, exceptions or conditions or qualifications should appear as closely as possible to the matters to which they refer. General conditions or qualifications applicable to various provisions in a clause should preferably be in the first sentence or sentences of the clause; specific conditions or qualifications and exceptions should be in the sentences to which they refer or immediately before or after those sentences; the following examples illustrate this point.

One of the commonest exceptions is that used to provide for occasions when the detailed drawings or the site conditions may require some local change from the general provisions of the specification. Thus –

> "*Except where otherwise specified or permitted by the Engineer* the maximum size of the aggregate used in concrete shall be 40 mm nominal size".

Such exceptions are essential (particularly in specifications) to take account of practical conditions in design and construction.

An example of the framing of a different sort of exception appears in GC/Works/1 where it is desired that the Contractor should take responsibility for the conditions under which work is to be carried out but should not be held responsible for the accuracy of conditions stated in Bills of Quantities, thus –

> "The Contractor shall be deemed to have satisfied himself as regards existing roads, railways, or other means of communication with and access to the Site, the contours thereof,... and generally to have obtained his own information on all matters affecting the execution of the Works and the prices tendered therefor *except* information given or referred to in the Bills of Quantities which is required to be given in accordance with the method of measurement expressed in the Bills of Quantities". (author's italics)

The following example illustrates a general qualification at the beginning of a clause –

> The following provisions shall apply only where an extension of time pursuant to Subclauses (1) and (2) hereof has been granted by the Engineer on account of any of the matters referred to in the Subsections (a) and (b) of this subclause.

The final example illustrates the practice of placing conditions and qualifications close to the matters to which they refer –

> Marine aggregates may be used if the contents of chloride salt in the aggregate, expressed as the equivalent anhydrous calcium chloride percentage by weight of the cement to be used in the concrete, does not exceed 1.0%. Where the proportion exceeds 0.1% by weight of cement, marine aggregates shall not be used with alumina cement nor for prestressed concrete. In addition, in concrete containing embedded metal, calcium chloride shall not be added in such proportion that the total anhydrous calcium chloride in the mixture plus

the equivalent value of anhydrous calcium chloride calculated from the chloride in the aggregate, exceeds 1.5% by weight of the cement.

2.24 Descriptions of Clauses etc.

A definition in the ICE Conditions says "All references herein to clauses are references to clauses numbered in the Conditions of Contract and not to those in any other document forming part of the Contract"; at various places in the text cross reference is made to 'clauses' and 'sub-clauses'. The FIDIC Conditions omit the definition, but cross references refer to clauses and sub-clauses. GC/Works/1 also has no definition but cross references in the text refer to Conditions and paragraphs within those Conditions. In other documents of the contract it is often necessary to cross refer to the Conditions of Contract and also to cross refer within the document and to other documents. Thus, when writing a specification it will be necessary to cross refer between various clauses of the Specification and, occasionally, to clauses of the Conditions of Contract; preambles to Bills of Quantities may cross refer to both the Specification and the Conditions of Contract. It is convenient (and will reduce verbosity) to adopt distinctive terms when cross referring to the various documents; the following terms can be employed, but it is important that, whichever terms are used, they are used consistently throughout each particular contract:

Conditions of Contract – Conditions and sub-Conditions
Specifications – Clauses and sub-clauses
Preambles to Payment Documents – paragraphs and sub-paragraphs
Bills of Quantities and other Payment Schedules – items

The foregoing terminology does not deal with other documents which may be found in engineering contracts such as formal agreements, subcontract forms, reports containing data, etc.; cross referencing to such documents is rare and, unless extensive references are required in a particular case, special terminology would not be justified. A full description such as "paragraph 3.2 of part 2 of volume 5 of the Site Investigation Report" may sometimes be necessary but verbosity in cross references distracts from the content of a sentence and thereby reduces readability. It is best to avoid cross references such as 'sub-sub-clause (a) of sub-clause 2 of clause 5 of section 3' and simply say Clause 3.5.2 (a).

CHAPTER 3

Arrangement and Form of Documents

3.1 Order and Logic

It has already been emphasised that documents must be read as a whole.
Unless drafting is carried out in accordance with a carefully formulated
plan it will be difficult to ensure that the various parts of the documents
are consistent with one another and to avoid ambiguities and discrepancies arising from interactions and overlapping between clauses in
different parts of the documents and between the various documents
making up the contract. Moreover, documents which contain haphazard
arrangements of clauses are difficult to read, with the result that some
of the provisions may be overlooked at a time when they should be
enforced; ease of readability of a document is a most important element
in effective administration of the contract by both the Contractor and
the Engineer.

When planning documentation for a contract, two fundamental principles should be adhered to:—

Each document and section of a document should be strictly limited
to the topics and matters intended to be dealt with in that document
or section.
The order of dealing with the various topics in each document should
conform with the natural logic appropriate to that document.

To achieve the greatest clarity it is desirable that each topic or matter
should be the subject of a separate clause which can stand independently

and not interact with other clauses. Although the draftsman should aim at such an arrangement, it is rarely possible to fully achieve it; practical requirements interact in a manner which does not admit their complete separation. A simple example of this problem is the specification of the formed surfaces of concrete, which depend upon the way in which the concrete is made and compacted and also upon the construction of the formwork. Although separate clauses are provided for formwork finishes and for concrete design and compaction, it is necessary to introduce under one or both headings a provision relating both concrete and formwork to the finished surface required.

Notwithstanding the practical problem, it is essential that the draftsman plan and draft on the principle of separation of topics and on restricting each document to the planned type of subject matter. As an example, the draftsman should avoid inserting pricing provisions or methods of measurement in the Specification; those provisions should be included in the Pricing Document where they can form part of a preamble or method of measurement drafted on consistent principles. Occasionally, for clarity and readability, it may be necessary to include in a specification clause a reference to some operation (most commonly in relation to remedial work) being carried out at the expense of the Contractor, but such references should be kept to an absolute minimum; other references to costs or payment should be avoided.

Natural logic in the arrangement of a document is not easily defined. It depends upon the nature of the document and upon the viewpoint of the draftsman; there is no definable best arrangement. In this context, the dictionary defines logic as a "chain of reasoning"; a number of consistent chains of reasoning may be equally applicable, providing that the presentation is such that the chain of reasoning is evident to the reader. The application of a consistent chain of reasoning or logic enables the various provisions to be made on the basis of general principles applying to the whole or parts of the works; consequently the draftsman can effectively subdivide the document into clauses each dealing with one topic (or one aspect of that topic), without overlooking the interaction of those topics, because he is able to relate them to the general principles and also to one another through the connecting logic.

Although the principles of order and logic are essential in the drafting of lengthy and complex documents, they apply equally to short or small contracts, even when only an exchange of letters is involved. Small contracts with short documentation almost invariably involve references to other documents (either explicitly or implicitly) and to technical practices; such basic requirements should, if possible, be made explicit. This requires careful thought and planning; even where such documentation and practices are implicit, the draftsman and the client should be aware of this fact and of its implications.

3.2 Preparing the Framework and Contents

The principles of order and logic require that the various parts of the documents be planned so as to ensure that:

(a) all requirements have been included and none have been overlooked;

(b) the separate documents forming the contract do not include contrary provisions;

(c) the provisions in each document do not interact in a manner which would result in conflict between them or in ambiguity or lack of clarity;

(d) the various provisions in each document reinforce one another so as to provide for all the requirements of each part of the Works.

In principle, the requirements of item (a) should be met by setting out the general requirements for the contract and, on that basis, deciding the character of the various documents and the requirements to be allocated to each of them. In practice, a number of standard documents exist (with clearly defined forms) and these standard documents are used in most contracts; the exceptions are generally contracts initiated by contractors, such as package deal offers, management contracting offers and the like. Even here standard forms rapidly develop, either within an organisation specialising in that type of offer or within the industry when such offers become common.

The convention adopted in most construction contracts is to subdivide the overall requirements for the work between four documents as follows –

(i) Conditions of Contract – dealing with the general requirements common to most contracts;

(ii) Specification – dealing with the general requirements of the Employer and the Engineer and with the technical and constructional requirements for materials and methods required by the design of the Works;

(iii) Method of Measurement – setting out the requirements for preparing the Bills of Quantities (where these are provided);

(iv) Bills of Quantities – provides the quantities of the various parts of the work and the accepted prices, to enable the final price to be calculated. In the case of manufacturing and erection or of fabricating contracts, an appropriate set of standard Conditions of Contract is usually available, but a special Payment Document (specific to the class of work) would be substituted for the Method of Measurement and the Bills of Quantities.

In building contracts (based upon a lump sum)it is common to combine the Specification and the Bills of Quantities and to use the Bills of Quantities for the purpose of pricing variations. As a result of this standardisation, the overall requirements have to be divided into classes appropriate to each standard document and the draftsman must be familiar with those standard documents to ensure that the necessary alterations and additions are made to provide comprehensive coverage for all the requirements of the particular contract.

Item (b) above will be satisfied if each document is limited to the topics specified for that class of document. In some cases, provisions are made in one document for the general aspects of a particular topic, while another document deals with provisions for those aspects of that topic which are specific to the particular contract. For example, although the ICE Conditions (in Condition 14) provides that the Contractor shall furnish a programme, it is often desirable to specify the type of programme and its contents in the general clauses of the Specification. Another related example is the provision in the ICE Conditions for completion in Sections; in this case it is essential to define, in the general clauses of the Specification, the precise extent of the work in each Section. Where all the provisions in connection with a particular topic cannot be limited to one document, the draftsman must explicitly take account of this in the notes on which his drafting is based, to ensure that the provisions in each document do not overlap; he should, where necessary, cross reference to provide compatibility between the provisions in each document. An example of such cross reference and further provision is a clause in a highway construction contract, which provides –

The programme required by Clause 14 of the Conditions of Contract shall be in the form of a time location bar chart covering all the main items of work, laid out in a format which will permit progress of the various items of the work to be indicated on it throughout the execution of the work.

The requirements of items (c) and (d) can best be met by formulating the contents and arrangement of the whole of the document before commencing drafting. This approach enables the draftsman to identify the interaction of the various parts of a large document before these interactions become obscured by the detailed drafting and to prepare a full contents list of all sections, clauses and sub-clauses. This contents list can then be circulated to other members of the team preparing the design and documents, for their comments.

3.3 Some Useful Conventions

Although there are no fixed rules applicable to the order and form of engineering contract documents, there are conventions which are usually, but not invariably, applied. These are:—

(1) Where an interpretation clause (containing definitions) is required, it is placed at the commencement of the document or of the section to which it applies. This convention applies to most legal documents, but not to Acts of Parliament, where the interpretation clause is placed at or near the end of the Act.

(2) Where the parties to the contract are defined in an interpretation clause, the definitions of the parties are placed at the commencement of the clause, the Employer being defined before the Contractor.

(3) General descriptions are given before descriptions or specifications of the details. If possible the details should follow closely behind the general description; however, this is not always appropriate and is likely to depend upon the form and style of the particular document.

(4) In description of work, substructures precede superstructures.

(5) Descriptions of work to be carried out, and of other duties, precede the details of payments to be made for such work and duties and these in turn precede particulars of the timing and methods of payment.

(6) Clauses concerning the settlement of disputes are placed at or near the end of the document.

In Bills of Quantities the order and form of the document is laid down in some detail in the Method of Measurement, which is normally a standard document applied to a particular type of work. Although Bills of Quantities are widely used throughout the world, the detailed Bill based upon a Standard Method of Measurement is peculiar to British practice. It is, however, becoming more common elsewhere.

3.4 Sections, Clauses, and Sub-clauses

Earlier legal documents were often written in long sentences or paragraphs with little or no subdivision within the document; they were extremely difficult to read and understand (and no doubt also to draft). Modern contract documents of any length are extensively subdivided, to improve readability and assist drafting. Even short documents are normally subdivided into paragraphs. The common subdivisions are:—

> Sections – intended to cover each of the main *topics* of the document.
> Clauses – intended to deal with each *subject* which is to be covered by the general topic.
> Sub-clauses – intended to deal with each of the *provisions* to be made in respect of each subject.

Some documents, such as the JCT Conditions and GC/Works/1, are not divided into Sections. In these documents it is generally more difficult to find particular clauses; also, the documents often appear to lack logic in the arrangement of clauses, to the detriment of readability. In the JCT Conditions, for example, Clause 15 deals with "Practical completion and defects liability" while Clause 21 deals with "Possession completion and postponement", and between these clauses are others dealing with "Assignment or sub-letting" and Insurances. Failure to group together clauses dealing with similar matters occurs also in GC/Works/1; examples (referred to by prefix C) are –

> C 27 – "Assignment or transfer of Contract" and C 30 – "Sub-letting", which are separated by C 28 – "Date for completion; Extension of time" and C 28A – "Partial possession before completion" and C 29 – "Liquidated damages"; C 32 – "Defects liability" is divorced from C 28 and C 28 a, despite the close relationship between the date for completion and the commencement of the "Maintenance period", resulting in failure to explicitly state that the Maintenance period commences at the date of completion.

By comparison, the ICE Conditions is divided into 25 sections, which improves readability and helps the draftsman to take account of the interrelation of similar matters.

A good example of sectional divisions in specification writing is the Highway Works Specification. This is divided into 26 sections plus appendices, each section covering a particular class of work, and designed to allow for substantial amendments and additions to meet the requirements of each individual contract.

Later chapters deal in detail with the form and arrangement of individual documents. It is sufficient here to emphasise the importance of dividing any lengthy document into sections, each section grouping together, in logical order, clauses dealing with the subjects covered by a particular topic.

The clause is the basic unit of most documents. It should be limited to one subject only and should be divided into subclauses; each subclause should comprise one paragraph and should be limited to dealing with one provision in respect of the subject of the clause. Generally, clauses are numbered and titled and included within the contents list, but sub-clauses are usually sub-numbered and not included in the contents list. However, where marginal notes or titles are used in the documents they are sometimes applied also to sub-clauses and are then included in the contents list; this practice is subject to wide variations, as can be seen by comparing different standard Conditions, such as the ICE Conditions and GC/Works/1.

The initial contents list prepared for drafting is best limited to clause headings; sub-clause descriptions are only appropriate where it is intended to use marginal notes and are, in any case, most easily dealt with after the sub-clauses have been drafted. As previously mentioned, an important advantage arising from the preparation of an initial contents list is that it can be circulated to colleagues dealing with other aspects of the project, to enable them to comment on the subjects of the clauses and to propose additional clauses relevant to their work. Such comments are most valuable in the early stages of drafting, in order to ensure that the clauses comprehensively cover all the subjects which need to be dealt with.

3.5 Marshalling Sentences

Marshalling of sentences is a device used in drafting to split up long sentences in order to improve clarity and readability and to reduce ambiguity. Although it is used extensively in legal documents and other explanatory writing (such as, for example, Plain Words by Sir Ernest Gowers), it is not a device used in literary writings; it appears to have no accepted word or title to describe it.

Marshalling has been referred to as "paragraphing" by some writers, although paragraphs are not involved when splitting up sentences; it has also been referred to as "tabulation", although no tables are involved. This latter description would be most unsatisfactory in engineering documents where tables are extensively used. In Chapter 2, Part 2.21, the description "item enumeration" is used as a convenient way of dealing with the use of the colon. The name "marshalling" has been adopted here because the process consists of splitting a sentence into phrases or elements which are then marshalled, by the use of layout, punctuation, and connecting words, to provide a form which enables each element to be read individually.

The following sentence from a specification is a simple example of how marshalling may be applied –

If the Contractor fails to lay the blinding layer immediately after the approval by the Engineer, the Contractor shall, at his own expense, excavate and trim any soil or rock of the founding stratum which may have deteriorated before the concrete blinding has been laid and shall clean up and remove any blown sand or other unsuitable material which may have entered the founding area and shall make up any resulting additional depth with concrete Class E.

In its marshalled form it would read –

If the Contractor fails to lay the blinding layer immediately after the approval by the Engineer, then the Contractor shall, at his own expense:
(a) clean up and remove any blown sand or other unsuitable material which may have entered the founding area; and
(b) excavate and trim any soil or rock of the founding stratum which may have deteroriated before the concrete blinding has been laid; and
(c) make up the resulting additional depth with concrete Class E.

The following is a more complex sentence, which could be used in marshalled form in a specification –

Piles shall comply with the general requirements for concrete and shall be supported, handled and pitched only as described in the Contract and stored (where so required) by placing on sufficient supports on firm ground to avoid damage by excessive bending and each pile shall be marked indelibly to show its identification, length and date of casting, and, in the case of prestressed concrete piles, shall be marked with the prestressing force applied.

In its marshalled form it would read –

Piles shall comply with the general requirements for concrete and they shall also:
- (a) be supported, handled and pitched, only as described in the Contract;
- (b) when stored, be placed on sufficient supports on firm ground to avoid damage by excessive bending;
- (c) be marked indelibly to show the identification number, length and date of casting on each pile;
- (d) be marked with the prestressing force applied to each prestressed pile.

In its simplest form, a marshalled sentence consists of introductory words followed by two or more elements. In more complex forms, some or all of the elements may be followed by short connecting words and by resuming words after the marshalled elements. Certain of the elements may also be followed by secondary elements. The following complex marshalled sentence is annotated to illustrate the relationships between all the constituent parts –

If, in connection with material dimensioned in the Contract (or ordered by the Engineer) in metric measure or dimensioned in imperial measure, the Contractor gives written notice to the Engineer setting out all the facts and stating that: } introductory words
  (a) he has used his best endeavours to procure such material in the specified measure; and ← } connecting words
  (b) he is unable to procure such material in the specified measure:
    (i) without additional cost; and/or ←
    (ii) without additional delay; or ←
    (iii) at all; } secondary elements

(c) such material is readily procurable in the alternative measure to dimensions (stated in the notice) which approximate to the dimensions in the specified measure;

then the Engineer, if he is satisfied that the Contractor has used his best endeavours (as stated in the notice) and has set out the full facts, shall (as soon as practicable after receipt of the notice) either: *intermediate resuming words*

    (d) instruct the Contractor to provide such material in the specified measure and allow the Contractor additional costs and time in accordance with the Contract; or ←————————   *connecting words*

    (e) instruct the Contractor to:

      (i) provide such material to the alternative dimensions set out in the notice; or ←————

      (ii)make some other variation which will avoid the use of material which has the specified dimensions;   *secondary elements*

or, if the Engineer is not so satisfied, he shall (as soon as practicable after receipt of the notice) inform the Contractor accordingly.   *final resuming words*

Primary elements are numbered (a), (b), (c), etc. and are not separately annotated.

This example illustrates the principles which should be observed when marshalling sentences. These are:—

1. The introductory words should be a phrase of substance which is common to all the following elements, up to any resuming words.
2. The introductory words should end in a colon. This aspect is discussed below under the heading of "Punctuating".
3. Each element should commence on a new line, should be indented and should be identified by a number or letter. An element should not commence with a capital letter (unless the first word is a proper noun) and should end in a semicolon.
4. Each element should be capable of being read continuously with the introductory words. An example is element (c) which can be read with the introductory words, as – "...setting out all the facts and stating that such material is readily procurable...". Where

there are intermediate resuming words then the elements following them should be capable of being read continuously with those resuming words rather than with the introductory words.

5. Where there are secondary elements, the initial words of the primary element should be treated as introductory words to the secondary elements and should end in a colon. Secondary elements should be further indented, should be identified by a lettering or a numbering system differing from the lettering or numbering system of the primary elements and should end in a semicolon.

6. Resuming words should be capable of being read as a continuation of the introductory words or of the previous resuming words. In the illustration the intermediate resuming words had clearly to be read with the introductory words as follows –
If, in connection with the material dimensioned.., the Contractor gives written notice... stating that..., then the Engineer, if he is satisfied that the Contractor....
Intermediate resuming words should end with a colon; final resuming words should end with a full stop. If there are no final resuming words, then the last element should end with a full stop instead of a semicolon.

Although there are differences between draftsmen in the use of the colon in marshalling, all the other principles are universally adopted by all draftsmen and should be strictly adhered to.

3.6 Listing

In engineering documents, particularly in specifications, it is often necessary to provide lists of various items or of drawings. These are not to be confused with marshalled sentences, although from casual inspection they appear to be similar; thus the list of principles for marshalling (set out above) has been written as 'listing' rather than 'marshalling'. The dividing line can sometimes be fine, in the case of a marshalled sentence which includes what appear to be a list of provisions which the Contractor or the Engineer has to make. The lists referred to here most commonly comprise physical items or objects, such as furniture or equipment or buildings or drawings, as in the following examples –

The following furniture and general equipment shall be provided for the offices and laboratories:—

3 No. typists' desks and adjustable type swivel typists chairs.
27 No. double pedestal desks and upholstered swivel type chairs.
24 No. lockable steel cupboards with full width shelving.
1 No. set of transparent plastic railway curves
2 No. steel straight edges.

A more complex list, taken from a specification –

The following items of laboratory equipment shall be provided and installed within the laboratory :—

1. Sieve Shaker and Sieves (sieves to B.S. 410 with lids and receivers)
   (a) Sieve shaker to take 6 No.sieves   1 No
   (b) 300 mm dia. sieves
       (i) 75, 63, 50, 37.5, 28, 20, 14, 10, 6.1, 6.3,
           and 5.0 mm (perforated plate)   2 sets
       (ii) 2.36 and 1.18 mm (woven mesh)   2 sets
       (iii) 600, 425, 300, 212, 150 and 53 micron
            (woven mesh)   2 sets
       (iv) 75 micron (woven mesh reinforced)   6 No
2. Riffle Boxes to BS 812
   (a) 50 mm nominal slot width   2 No
   (b) 375 mm nominal slot width   1 No

These examples illustrate the principles which should be observed when listing. They are:—

(1) The introductory words do not have to be read with each item; they stand on their own.
(2) The introductory words end in :—
(3) Each item is to be indented and may be numbered; numbering is not essential.
(4) Where there are sub-items they should be further indented and preferably numbered or lettered within each item.
(5) Each item should commence with a capital letter and end in a full stop.

3.7 Punctuating and Numbering for Marshalling and Listing

There does not appear to be any universal practice for punctuating the end of the introductory words in marshalled sentences. Robinson (in The Principles of Drafting) recommends that a dash be used at the conclusion of introductory words; this is the most common usage in legislation, although even in such documents the colon or colon dash is occasionally used. Marshalling is extensively used in standard forms of Conditions of Contract for engineering works. Examination of the various national and international standard forms for civil engineering, for mechanical and electrical engineering and for building shows that all three alternatives (i.e. colon or dash or colon dash) are in use, sometimes all three in different parts of the same document. The recommendation given above that a colon be used at the conclusion of introductory words is based partly on the fact that most engineering documents today (other than standard forms) are reproduced from typescript prepared on typewriters or on word processors. These machines can type hyphens (one space in length) but not dashes (two or three spaces in length), which is an unsatisfactory substitution when used at the end of a line, although it has a satisfactory appearance when used with a colon. As it is necessary to distinguish between marshalling and listing by using the colon dash with the introductory words to a list, the colon should be used with the introductory words when marshalling sentences.

Occasionally a draftsman uses the comma in place of the semicolon at the end of an element in a marshalled sentence. This practice is sufficiently rare for it to be ignored; the semicolon should always be used to conclude each element.

In marshalled sentences, each primary element is always identified by a lower case letter, usually in brackets; secondary elements are identified by lower case roman figures, also in brackets. In the case of lists, there appears to be a wide variation in practice; in some cases items are numbered with Arabic numerals or occasionally with Roman numerals, in other cases there is no separate identification (other than indenting and starting a new line for each item), while in yet other cases lower or upper case letters are used, particularly where there are sub-items in the list. For uniformity, it is suggested that each item should have an Arabic number, sub-items being distinguished by lower case letters and sub-sub-items by small Roman numerals. Examples of all the foregoing systems of numbering or lettering for identification are used in the illustrations and examples of marshalling and listing given above.

Marshalling and listing, as drafting devices, should not be confused with the sub-division of sub-clauses into sub-sub-clauses or sub-sub-paragraphs. In this case the sub-sub-clauses or sub-sub-paragraphs are not introduced by introductory words, as is the case in marshalling or listing, and each stands on its own within the sub-clause in the same way as each sub-clause stands on its own within the main clause.

3.8 Annexures, Schedules, and Appendices

It is sometimes necessary to include, within a particular clause in a document, a list of items or a description or some other requirement of considerable length which, if inserted into what would otherwise be a relatively short clause, would obscure the meaning or intention of that clause. The solution is to remove the lengthy description and to include it in an annexure to the clause, which can be placed at the end of the section of the document concerned or at the end of the document. This is particularly useful in standard documents, where such descriptions may apply only to particular contracts and which are not, therefore, able to be included within the standard. Such annexures are often referred to as Schedules and are very common in statutes and other legislative documents. However, an annexure to a clause is exclusive to that clause, whereas a Schedule may be independent but referred to in a number of different clauses. The nomenclature is not very important. A good example of the use of Annexures or Schedules occurs in the Roads and Bridges Specification, which provided that –

> "The Contractor shall provide, maintain and, except those listed in accordance with sub-clause 5 below, remove on completion of the Works the offices and testing laboratories for use of the Engineer, including their contents, as described in the Schedule to this Clause."

In any contract in which this standard form of Specification was included, the draftsman had to include a Schedule setting out the requirements for the Engineer's office.

Where the document requires one party to complete a form at some stage in the Contract, then the "form" can become an annexure to a clause which provides for the completion of that form.

Occasionally it is necessary to provide information to accompany a document, which, although it does not form part of the subject matter of the document, is most conveniently attached to it. An example of this is Clause 14(5) of the ICE Conditions which requires the Engineer to provide the Contractor with relevant design criteria for the Permanent Works as may be necessary to enable the Contractor to design the

Temporary Works. If this information is to be given at the tender stage then it is most conveniently provided in the form of an Appendix to the Specification; it cannot conveniently be appended to the ICE Conditions (which are usually only incorporated in the contract by reference) and it may not form part of the contents of the Specification.

As an example, a large contract had four appendices as follow:—

Appendix 1 – Note of principal matters for liaison between the Employer and the Statutory undertaker
Appendix 2 – Design assumptions
Appendix 3 – Programme Schedule
Appendix 4 – List of approved manufacturers of protective coating systems.

3.9 Definitions – Types and Usage

The use of definitions is another device adopted in drafting to achieve clarity and to reduce ambiguity. Their skilful use assists precision of description, helps in avoiding changes of meaning when using the same words and can result in a significant reduction in the length of a document as well as improving readability.

The simplest form of definition is of the name of one of the parties to a contract, such as the Employer or the Contractor. This type of definition has the double advantage of ensuring the avoidance of typographical errors in repeating the name and of shortening the drafting by avoiding repetition of lengthy names. It can also be used to avoid lengthy repetition of other names in the text, either by use of a "nickname" or of a descriptive word or phrase as, for example, "Ministry of Agriculture, Fisheries and Food (referred to as MAFF)" or "Imperial Chemicals Industries Limited (referred to as the Paint Manufacturer)". A variation of this type of definition occurs in the following statement, used in a specification –

References to clauses of the Civil Engineering Specification for the Water Industry are prefixed WI.

Where the meaning of a word is not clearly established in the language habits of any section of the readers of the document, a dictionary type definition (known as a lexical definition) is required. The meaning of the term "sketch plate", although known to steel fabricators, does not commonly form part of the vocabulary of structural designers or of engineers concerned mainly with erection of steelwork. A lexical definition is, therefore, required as follows –

> "sketch plate" means a plate whose shape cannot adequately be described in words and which requires a sketch to define that shape.

Often the lexical meaning or meanings of a word or phrase are well known, but the extent of their application may be interpreted in different ways. In this case, a definition of the extent of application of the term is required, such as –

> ' "Constructional Plant" means all appliances or things of whatsoever nature required in or about construction completion and maintenance of the Works but does not include materials or other things intended to form or forming part of the Permanent Works.' (from ICE Conditions)

Another form of this type of definition does not seek to define the term, but merely to ensure that certain aspects are covered. The definition uses the word "includes" rather than "means". It is most useful for ensuring that the expressions used in the document are comprehensive and enables this to done without undue repetition, as the following examples (from the Law of Property Act 1925 and from GC/Works/1 respectively) illustrate –

> "The singular includes the plural and vice versa"

and a more complicated provision

> "The expression 'loss of property' includes damage to property loss of profit and loss of use."

The Law of Property Act 1925 is a source of many legal definitions which, because they have the force of law, are applicable without being repeated in a document. Most of these definitions are not appropriate to engineering documents; they are intended for buying and selling property. However, Section 61 (quoted in Chapter 2 Part 2.13) provides definitions of "Month", "Person", singular and plural, and masculine and feminine. Its provisions are, of course, only applicable in the UK, although many Commonwealth countries have a similar Act which incorporates these definitions.

In some cases it is convenient to split a definition between two of the documents forming a contract, the general part of the definition being given in a standard document while the detailed aspects are defined in a document specific to the work concerned. The most common examples are the definitions of the Works and of the Site. The general requirements are defined in standard Conditions of Contract; specific requirements applicable to the particular contract are most conveniently provided in the Specification under the heading of "Description and Extent of Works" and "Location and Extent of Site" respectively. These specification clauses should be treated as additional paragraphs to the Conditions of Contract definitions and may commence with the words "The Works comprise..." or "The Site comprises..." The definitions of the Site are, of course, only applicable to construction contracts or to plant contracts with erection; they will not apply to contracts for the supply of mechanical and electrical plant. The subject is dealt with in detail in Chapter 5 Parts 5.7.1 and 5.7.2.

When framing definitions the draftsman should avoid repeating the words of the dictionary; the definition in the document should be limited to supplementing or adjusting the dictionary meaning in order to provide for the requirements of the document being drafted.

3.10 Interpretation Clauses

Where there are a number of definitions applicable throughout the document, it is usual to group them into an interpretation clause. As mentioned in Part 3.3, it is customary to put the general definitions in an interpretation clause at the beginning of the document; definitions applicable only to one section of the document are, however, usually retained within that section.

In order to guard against the risk that a defined word in the text of the document may be inadvertently, but obviously, used in a sense other than the one defined, it is usual to commence an interpretation clause with words such as –

"The following expressions shall, unless the context otherwise requires, have the meanings respectively assigned to them :—"

A list of definitions follows these opening words, the rules for punctuation being as generally set out in Parts 3.6 and 3.7 above. However, the expression which is being defined is, in each case, put in inverted commas and the words forming the expression commence with capital letters; this device clearly identifies the expression, the capital letters (but not the inverted commas) being retained throughout the document whenever the defined expression is used with its defined meaning. If the same expression is used without the defined meaning, then capital letters are not used. An example of this is the expression "the Contractor" which is defined to mean the particular contractor to whom the contract has been awarded. When some other contractor is referred to then the lower case is used, as in the statement –

> The Contractor shall liaise with the contractor for contract No 1 and shall agree with him the dates on which it is expected that the permanent drainage will be connected and shall make arrangements for discharge of drainage to any temporary drains or pumping sumps operated by the contractor for contract No 1.

Definitions which form part of a section or a clause in a document are not usually preceded by the qualifying words used in interpretation clauses; it is relatively easy to check a section or clause to ensure that no expressions used in the text are inconsistent with the definitions.

It is often convenient to form a definition by putting the defined expression in brackets after its description or meaning, as in the following example –

> The rates and prices take account of the levels and incidence at the date for the return of tenders (referred to as "the relevant date") of the taxes levies and contributions which are by law payable...

The bracketed definition is provided on the first occasion on which the expression is used and it is usual to omit the words "referred to" (although they are sometimes used), these words having become implicit as a result of the continued usage of this style of definition. This particularly applies to definitions of names, e.g. –

> Unplasticised polyvinyl chloride (UPVC) pipes complying with...

thus avoiding extending the interpretation clauses and reducing the length of the document.

3.11 Definitions to be Avoided

In theory, words can be given any meaning that one wishes; it is quite possible to define white as meaning black. As Humpty Dumpty said in Alice in Wonderland "words mean what I say they mean". Although unnatural meanings can be tempting, particularly when one is attempting to hurriedly amend a document, they form a barrier to communication. Notwithstanding the definition, the reader of the text is likely to interpret the word in accordance with its ordinary meaning, giving rise to mis-understandings and disputes. Moreover, the draftsman is liable to forget the unnatural definition at some point in the drafting of a lengthy document, with the resulting risk of ambiguity. Dickerson refers to the ultimate in Humpty Dumpty definitions which occurred in Bills which came before the U.S. Congress and which provided that, for their purpose the term "September 16th 1940" meant "June 27th, 1950"; one can imagine the mental gymnastics necessary to understand such legislation. He also draws attention to an English statute which provided that –

'Whenever the words "cows" occurs in this Act it shall be construed to include horses, mules, asses, sheep and goats'.

In an engineering context, one can foresee considerable difficulty with the definition –

Engineer means ...., chartered accountant of Holborn, London.

Language habits invest words with meaning which cannot be shrugged off; defining a chartered accountant as an engineer does not give any confidence in his ability to settle engineering disputes in a contract. The problem is best summed up by Robinson who says (in his book "Drafting") –

"Because it is not possible to cancel the ingrained emotion of a word by an announcement, words should be defined by reference to a meaning that they can bear".

Occasionally it becomes necessary to give a word an unnatural meaning; the effect can be minimised by the use of the "as if " technique. An example relates to the term "water/cement ratio" which is used in specifications for concrete. In modern practice a mixture of cement and pulverised fuel ash (PFA) may be used in place of pure cement for part of the concrete for a project. It is tempting to say –

"PFA means cement where the context requires."

It is far better to give a fuller definition and say –

'In the expression "water/cement ratio", the weight of any specified combination of cement and PFA shall be considered as if it were the weight of cement.'

An alternative is to use two definitions as follows –

"Cementitious content" of concrete shall mean either the cement content or the content of cement and PFA combined. In the expression "water/cement ratio" the words cementitious content shall be substituted for cement.

When framing definitions it is necessary to avoid including substantive provisions. Dickerson gives an example, from draft zoning regulations, of the problems that can arise in this connection. The proposed draft definition was –

' "parking space" means a space, no smaller than 9 ft by 20 ft, for the offstreet parking of one motor vehicle.'

Unfortunately, this allows a householder to locate a 9 ft by 19 ft parking bay in front of his house, contrary to the intention of the regulations. The problem was overcome by omitting the substantive provision from the definition thus –

'The term "parking space" means a space for the offstreet parking of one motor vehicle. No parking space may be less than 9 ft by 20 ft.'

Similar problems could occur in a specification definition which said –

Cementitious content means a mixture of 70% cement and 30% PFA.

Confusion would arise if, at any time, it became necessary to specify some other mix proportion, i.e. the proportions (the substantive provision) should be omitted from the definition.

In a definition, the word "means" limits the meaning of the defined term to that given in the definition; the word "includes" widens the meaning to include matters which might not formerly be included. The two words are, therefore, incompatible and the term "means and includes" cannot be used, i.e. we cannot, for example say –

"Rock" means and includes hard material on its natural bed and boulders exceeding 0.2 cubic metres in volume.

It is equally inappropriate to say –

"Rock" means hard material on its natural bed and includes boulders exceeding 0.2 cubic metres in volume.

This problem can be solved by avoiding the word "includes" and adding a separate provision, as follows –

> "Rock" means hard material on its natural bed. Hard material in boulders exceeding 0.2 cubic metres in volume shall be deemed to fall within the definition of rock.

An alternative method is illustrated by the changes which were made in the Roads and Bridges Specification. In the 1969 edition there is a definition –

> ' "Unsuitable material" shall mean other than suitable material and shall include: .... '

this is followed by a list of materials which are included. However, in the 1976 edition this was revised to state –

> ' "Unsuitable material" shall mean other than suitable material and shall comprise: ..... '

this is followed by a list of materials (the same list as the 1969 edition), i.e. the word "comprise" limits the meaning to those items listed and thus is compatible with the word "means" and does not extend beyond it.

3.12 One-off Definitions

It is obviously inappropriate to include in any interpretation clause a definition of a word or phrase which is used only once or twice in the text of the document. The reader may be unaware that the term has been defined and it will not improve readability if he has to search for definitions of terms which are used only occasionally. It is better to give a full description in the place where the term is used. In some cases, the description of a term may be lengthy and complicated and may interrupt the flow of the language and obscure the meaning of the provision in which it occurs. In such a case the term should be defined immediately before or after the paragraph in which it is used. A good example of this occurs in the IEE Conditions clause for variation in labour taxes, which says –

> 'If, as a result of the coming into effect after the date of the tender of any change in the level or any incidence of any labour-tax matter,

including the imposition of any new such matter, or the abolition of any such matter previously existing the cost to the Contractor of performing his operations under the Contract shall be increased or reduced, the amount of such increase or reduction shall be added to or deducted from the Contract price as the case may be.'

'In this sub-clause "labour-tax matter" means any tax levy or contribution (including National Insurance contributions but excluding tax and any levy payable under the Industrial Training Act, 1964) which is by law payable by the Contractor in respect of labour and any premiums and refunds which are by law payable to the Contractor in respect of labour.'

The use of such a one-off definition is common in legislation and can be very valuable in engineering documentation.

3.13 Mathematical Formulae, Tables, and Forms

Mathematical formulae are not very common in contract documents, as they are mainly concerned with quantitative relationships, whereas the documents deal with responsibilities and duties, with methods of construction and with the duties required of mechanical and electrical plant. In Conditions of Contract, formulae generally appear only in connection with payment adjustments, such as the BEAMA Formula for adjusting for inflation in the UK in mechanical and electrical plant supply contracts, and in the Baxter and Osborne formulae, for making similar adjustments in civil engineering and in building work respectively. Formulae have also been used for determining payments to be made for engine overhauls and also for payments to be made in respect of loss of productivity due to industrial disputes. In specifications, formulae are usually employed only in connection with laboratory testing and quality control; other matters concerning quantitative relationships are usually dealt with by means of tables.

When defining the symbols in a formula it is usual to use the sign $=$ instead of the word "means"; the definition should otherwise be treated similarly to all other definitions and particular care should be taken to avoid including substantive provisions in those definitions. Such substantive provisions should be incorporated in the formula or in limitations applied to the formula.

Tables are very common in specifications and similar documents; they have the advantage that the figures in them do not have to be related by continuous mathematical functions and that the highest and lowest values in the tables automatically limit the extent of applicability of the table. Because of the wide variety of data and relationships which may be assembled in tables, there are very few general rules applicable to them. However, it is important that every table be numbered and titled (preferably in a manner which is easily related to paragraph numbers in the text) and that references are to the table number and not by phrases such as "the accompanying table". The text of a document often changes in the course of drafting and table references may become incomprehensible unless clearly linked to the text by a number or a letter. Where there are a number of columns in a table, it is desirable to number the columns to enable textual references to be clear and unambiguous. Column and line description must be carefully drafted to ensure clarity, particularly in relation to such phrases as "not less than" or "not greater than", and explanatory notes are required where the possibility of interpretation arises. Where the column headings are very lengthy it may sometimes be necessary to set them out separately (either above or below the table) in a note, with numbers related to the column numbers. This is preferable to condensing the headings to an extent which may give rise to difficulty in understanding or to ambiguity.

It is sometimes necessary to provide that some information (such as a certificate or a declaration) shall be set out in a particular way or shall involve certain specified wording. This is done by annexing a standard form to the document. Such a form should preferably be on a separate sheet and should have a heading describing its purpose; if there are several such forms it is worthwhile giving each form a number for cross-referencing in the text, in a similar manner to table numbers. Blanks left for the insertion of names or descriptions or other information should have dotted lines on which the information may be written and, for clarity, they are often provided with a description, in brackets, of the information which is to be inserted in the blank space.

3.14 Document Layout

Documents can and have been written as one continuous text, without punctuation and with only the occasional full stop, if any. This old style of legal drafting is practically unreadable. Apart from punctuation, one of the most important devices for improving readability is good layout of the typescript, which provides headings, numbering and spacing to break up the solid mass of the document text.

Breaking the document text into Sections (and in some cases grouping the Sections into Parts) provides the equivalent of chapter headings. Providing numbering and sub-headings to each clause and the further numbering of sub-clauses divides the typescript and reduces that weariness in the reader which results from reading a continuous text. Adequate spacing between clauses and between sub-clauses also helps.

Lawyers and some engineer draftsmen favour the use of marginal notes for clauses, rather than sub-headings; they are also used in some documents to describe sub-clauses as well as clauses (as in the ICE Conditions). Most standard Conditions of Contract are printed with marginal notes, but specifications are more commonly printed with sub-headings.

Marginal notes have the advantage that they can be written after the document has been drafted, so enabling the draftsman to select the most apposite clause description. They are less effective in providing a subdivision of the typescript, but this disadvantage can be overcome by variation in the size and weight of the typeface used for the marginal notes. There are a wide variety of typefaces available for printing standard documents, but this variety is not available with the methods usually adopted for printing non-standard documents such as specifications. For this reason, sub-headings are more commonly used than marginal notes for this type of document. There are, of course, some hybrids combining headings with marginal notes; an interesting example of this is the FIDIC M & E Conditions which has both marginal sub-headings overlapping the typescript and marginal notes (this has changed in the 1987 edition). An example of the use of sub-headings in a printed document is the Highway Works Specification; it is unusual in being printed in half page width columns, rather than full page width.

Clauses in lawyers documents are generally numbered consecutively from the commencement of the document, sub-clauses being numbered consecutively within each clause and this system is usually adopted in standard Conditions of Contract, whether they are sub-divided into sections or not. The system in short documents (such as Instructions for Tendering) is usually limited to consecutive numbering of paragraphs, but in longer documents (such as Specifications) which are divided into sections, one of three alternative systems of numbering is usually adopted:—

(a) Consecutive numbering commencing at the beginning and continuing for each clause, notwithstanding the division into sections;

(b) giving each section a hundreds digit and numbering the clauses consecutively in each Section from that digit, e.g.in Section 300 – Earthworks, the first clause is numbered 301;

(c) sections numbered consecutively and the clauses numbered by the decimal system within each section, sub-clauses and sub-sub-clauses being numbered by further decimal additions within the clause numbers, e.g. in 'Section 3.0 – Earthworks' the third sub-clause of the first clause would be numbered 3.1.3

Method (b) has the advantage that additional clauses can be added within a section without disturbing the numbering system. It is particularly useful when amending a standard specification or adapting an existing specification for a new project.

Standard documents such as Conditions of Contract have usually been printed by letterpress, i.e. set up by compositors, with a very wide variety of typefaces. The number of copies produced often runs into many thousands or tens of thousands, whereas the number of copies required for an individual contract may number perhaps fifty. Even with this small number it was previously economical to use letterpress printing, but with the advent of small offset litho machines using paper plates, letterpress printing for individual contracts has become unusual. Today most of the documents for individual contracts are reproduced from typewritten originals, by offset litho. The modern office typewriter, either electric or electronic, often has interchangeable golf balls or daisy wheels which allow some limited variations in typeface. It is not usually feasible to combine bolder headings with ordinary type on the same machine when typing a document and this limits freedom of layout as compared with the variety offered by letterpress. However, many documents today are being produced on word processors, which do allow a variety of typefaces to be used in the same document, although this variety is rarely as great as that available with letterpress. An example of this is the use in letterpress printing of a different typeface (often smaller and heavier) in marginal notes. This option is rarely satisfactorily available on word processors, although the situation is changing with the advent of desktop publishing programs.

Another matter affecting layout is the decision whether to print on one or both sides of each sheet. This particularly affects the use of marginal notes, which usually require the margins to vary on alternate pages, the note margin being wider on the right when double sided printing is used. Although double sided printing can be undertaken on small litho machines, it is usually more convenient to print one side only and this is the commonest format when this method of printing is adopted. In many documents, such as Bills of Quantities, there is an advantage in single side printing as it provides a blank page on the left hand side on which notes and comparisons can be made when considering tenders or during the administration of a contract.

When the final draft of the document has been typed, the typist or word processing operator should number all the pages and prepare a contents list of clauses. Although many legal documents do not have contents lists, in contract documents these are essential to facilitate reference during administration of the contract.

The layout adopted must obviously be related to the means used for reproduction. Some large firms of engineers have adopted house styles for particular kinds of document, but these are not always adhered to, the layout style being affected by the needs of particular projects, including clients' wishes and the customs in the various parts of the world where projects are carried out.

Where typesetting is carried out by a commercial printer (with a large variety of typefaces available), he will be able to advise on layout and typography. Where reproduction is from typed or word processed originals and where there is no fully developed house style, the draftsman should provide for a simple layout which will enhance readability and clarity. Such a layout could be based upon the following outline scheme:—

(a) Section headings centralised and in bold capitals.
(b) Clause numbers and headings commencing at left hand margin in bold type with words commencing in capitals.
(c) Both left and right hand margins 25 mm wide for A4 paper and, if possible, the right hand margin justified, i.e. the ends of each line being aligned at the right hand margin (this can only be done on word processors or on certain electronic typewriters).
(d) Line spacing to be:
    (i) generally one and a half space;
    (ii) between sub-clauses and between clause heading and first sub-clause – 3 spaces;
    (iii) between end of clause and heading of next clause – 4 spaces.

(e) Printing one side.

When bold type face is not available for headings or its use is unduly onerous (such as changing a golf ball solely for typing a heading), the alternative is underlining of ordinary type face.

Binding of document volumes often depends upon availability of binding resources. The simplest binding available today is the plastic comb or spiral binding but this is not sufficiently durable for hard use; it is, however, suitable for tendering purposes. Copies of documents which are required for contract administration or for use at site should have a more durable binding such as stapling combined with heavy self adhesive plastic backed cover tapes. Where there are a number of contracts within a project it is helpful if the colours of covers and backing strips are varied with each contract, so that the document volumes for a particular contract are easily identified in the bookshelf.

3.15 Incorporation of Standard and Existing Documents

In practice, a large proportion of engineering documents consist either of standard documents amended to suit the contract or previous contract documents cannibalised and added to or amended. In order to avoid incongruities and ambiguities, the drafting and the punctuation style of the original may have to be adopted.

Conditions of Contract are usually standard printed documents which are incorporated by reference, subject to amendents, i.e. the printed standard form is not bound into the documents but is referred to in a paragraph which makes it subject to amendments which follow the reference. A typical reference paragraph (referring to the FIDIC Civils Conditions) would be –

> The Conditions of Contract shall be the PART 1 – General Conditions incorporated in the CONDITIONS OF CONTRACT (INTERNATIONAL) FOR WORKS OF CIVIL ENGINEERING CONSTRUCTION (4th Edition: September 1987) published by the Fédération Internationale des Ingénieurs-Conseils (FIDIC) as modified and added to by the Part II – Conditions of Particular Application set out herein.

Although the printed Conditions generally have marginal notes, this does not apply to the amendments, which are numbered in accordance with the Conditions being amended and have sub-headings corresponding to the main marginal headings. Additional Conditions are added after the amendments and are numbered after the last clause number of the printed Conditions (most printed Conditions make provision for this).

Very occasionally, the printed Conditions are themselves amended by the cut and paste technique and copied into the documents, but this method is more commonly applied to produce working documents for administration purposes. A typical example of a 'cut and paste' clause is shown in fig.1 where the typewritten additions are easily distinguished from the main printed wording.

47.   The whole of the materials, plant and labour to be provided by the   Rate of
Contractor under Clause 5 hereof and the mode, manner and speed of execution   progress
and maintenance of the works are to be of a kind and conducted in a manner
approved by the Engineer.   Should the rate of progress of the works or any
part thereof be at any time in the opinion of the Engineer too slow to ensure the
completion of the Works or any Section thereof within
the period limited in that behalf as aforesaid or
any extension thereof, then the Engineer shall so notify the Contractor in writing
and the Contractor shall thereupon take such steps as the Contractor may think
necessary and the Engineer may approve such steps to expedite progress, so as
to complete the Works or any Section thereof within
the period limited in that behalf as aforesaid or any
extension thereof.   If the work is not being carried on by day and by night and
the Contractor shall request permission to work by night as well as by day, then if
the Engineer shall grant such permission the Contractor shall not be entitled to
any additional payment for so doing but if such permission shall be refused and
there shall be no equivalent practicable method of expediting the progress of the
work the period for completion of the Works or any
Section thereof shall be extended by the Engineer
by such period as is solely attributable to such refusal.   All work at night shall
be carried out without unreasonable noise and disturbance.   The Contractor
shall indemnify the Council from and against any liability for damages on account
of noise or other disturbance created while or in carrying out the work and from
and against all claims, demands, proceedings, damages, costs, charges and
expenses whatsoever in regard or in relation to such liability.

Figure 1

Standard printed Conditions are negotiated between professional organisations and contractors in each branch of the engineering and construction industry. Staff dealing with tenders and with administration become familiar with the standard conditions and find it easier to read the amendments separately, rather than search for them in a reprinted edition of the standard form. For this reason, it is preferable to write out the amendments in the usual manner rather than to amend and reprint the standard document.

Apart from Conditions of Contract, the other main type of printed standard form is limited to the civil engineering and building industries and comprises the Method of Measurement, upon which the Bills of Quantities are based. Amendments to these documents are usually incorporated in the Preamble to the Bills; the subject is dealt with in publications concerning preparation of Bills of Quantities and does not require repetition here.

Some industries have standard printed forms of specifications. One of the best examples of this in the UK is the Highway Works Specification of the Department of Transport; this specification is usually incorporated by reference, using a standard clause published by the Department.

Most specifications are prepared by using clauses for general requirements and materials and workmanship which are taken from a previous document and modified to suit; it may almost be said that the art of specification writing is mainly the skilful practice of plagiarism. However, the art of drafting this type of document lies in ensuring that all the clauses cover the specific requirements of the work in hand, that the necessary additional clauses dealing with the special requirement of the particular contract are included, and that the requirements of the Works are comprehensively covered. Skill in drafting, experience in checking and the importance of a logical arrangement cannot be over emphasised; the subject is more fully dealt with in Chapter 5.

CHAPTER 4

Conditions of Contract

4.1 Purpose of Conditions of Contract

The document referred to in engineering contracts as "Conditions of Contract" is intended to regulate the relationships between the parties to the contract; it defines the parties and their responsibilities to each other as well as their responsibilities for various aspects of the contract. The liabilities of each of the parties and the respective risks to be taken by them are usually implicit in the defined responsibilities, but because of the complexity of the usual form of Conditions of Contract, it is preferable for those liabilities to be explicitly defined.

The relationships defined in the Conditions are subject to the provisions of contract law and the documents must be drafted in accordance with those provisions. Engineers concerned with the drafting of Conditions, or amendments to them, should have a good knowledge of contract law; when in doubt they should take legal advice on the subject. The legal aspects relating to some of the requirements of the documents (such as the law related to bankruptcy and the regulations governing VAT) are beyond the expertise of most engineers and require legal advice and drafting. When a new standard form of Conditions is drafted, it is common for the draft to be vetted by a lawyer specialising in construction contracts; similar vetting is sometimes required when substantial amendments are made to existing forms. It is essential that lawyers who check engineering contracts should have specialist experience in the subject; a general lawyer without this specialist experience is unlikely to be able to provide sound advice on the subject.

Because of the nature of engineering contracts, particularly those involving a substantial amount of site work, there is a need for administration, supervision and inspection throughout the period of the contract. In the British system, which is generally used throughout the Commonwealth (and in a modified form in North America), and elsewhere in the form of the FIDIC Conditions, a professional engineer or consultant is named in the interpretation clause and is usually entirely responsible for administration of the contract and for deciding disputed points in a manner which is fair to all the parties to the contract. The powers and responsibilities of the Engineer are a fundamental part of the contract and their proper definition and description are important items in the drafting of the contract documents.

Some oil companies working in the North Sea have substituted "the Company Representative" for "the Engineer", in order to provide direct control rather than working through an independent engineer. If the work is to proceed smoothly and disputes are to be minimized, the Company Representative needs to carry out his duties in a similar manner to the Engineer. In many contracts for public authorities, the Engineer is a member of the Authority's staff. Although the Engineer employed by a public authority might or might not have greater independence than the Company Representative, the difference is likely to be one of degree rather than fundamental; there should be little difficulty in adjusting standard Conditions to suit the actual duties and responsibilities.

Other contract systems do not necessarily provide for the role played by the Engineer in the British system. In the Swedish form of contract the Engineer's function is limited to inspection to ensure compliance with the technical requirements of the contract; the Engineer may undertake day to day supervision on behalf of the Employer but, in this respect, does not have the standing of the British Engineer. General administration of the contract is undertaken by the Employer who will settle disputed matters by negotiation with the Contractor. In many countries on the continent of Europe, contract law is related to or based upon the Code Napoleon; Conditions of Contract are either very limited or non-existent, reliance being placed upon the 10 year legal responsibilities of the Contractor. Elsewhere, contract forms may vary from a simple verbal agreement (as referred in the quotation from a Japanese contractor at the beginning of Chapter 3) to systems based largely on the British Conditions, particularly since the advent of the FIDIC Conditions. Although many aspects of the British system of engineering contracts which are incorporated in the FIDIC Conditions can be deleted or substantially modified by the "Conditions of Particular Application" which are added to the general document, the basic principles are generally accepted by the users of these contract forms.

Contracts for the construction of buildings may be considered in the same category as civil engineering contracts. Many buildings are constructed under the ICE Conditions; the government contract form GC/Works/1 covers both building and civil engineering works. Building contracts in the British system introduce two new characters to the draftsman, namely the Architect and the Quantity Surveyor, who jointly perform the functions of the Engineer in an engineering contract. To assist in supervision and settlement of disputes, other professionals are sometimes introduced into engineering contracts. In a contract involving a review clause, professional auditors were included in the Conditions of Contract, with the duty of preparing audit reports of the Contractor's financial operations.

Most contracts provide for arbitration in case of disputes which are not resolved under the general contract provisions. This is intended to avoid expensive and protracted litigation, but whether it succeeds in reducing the expense or time depends upon the terms laid down for arbitration and the method by which it is conducted.

When drafting Conditions or amendments to them, it should be borne in mind that the Contractor plays a dynamic role in the contract, while the Employer's role is passive, i.e. the Contractor initiates actions and claims while the Employer deals with matters after the event. The balance between the parties is maintained by providing the Employer and the Engineer with powers under the contract which will enable them to act effectively against any detrimental actions by the Contractor.

4.2 Fundamental Logic

Consideration of the purpose of Conditions of Contract in conjunction with the requirements of the law of contract enables a logical framework to be devised for the drafting them. Fundamentally, the document can be divided into four parts as follows:—

Definition of parties to the contract
Description of contract
Duties of Contractor
Payment by Employer

These divisions have been placed in serial order, i.e. they are in the order in which the various actions are presumed to take place: thus, the description of the works follows the agreement of the parties to form a contract, the Contractor's duties (or the actions to be taken by him) follows the description of the contract (which says what has to be done) and payment follows the Contractor's action in carrying out his part of the Contract. Arranging the sections and clauses in serial order is an important part of the logical framework for drafting this type of document. Some items do not have any definable serial order and, in that case, should be grouped together so that the group concerned is arranged in serial order in relation to other clauses and groups of clauses.

The fundamental logical arrangement given above is, of course, too simple to cover even the basic requirements for drafting Conditions for engineering. Without entering into any consideration of the content of the contract, the items in the basic logical arrangement need to be expanded to include the basic provisions for the relationships between the parties. The expanded basic fundamental logic is as follows:—

A– Definition of the parties to the contract
B– Duties of the Engineer and his staff
C– Description of Works
D– Duties and responsibilities of the Contractor
E– Duties and powers of the Employer
F– Payments by the Employer
G– Settlement of disputes

Because the duties and responsibilities of the Contractor and of the Employer interact, there will be overlapping between items D and E. Moreover, many of the provisions within the various items may have no obvious serial order (particularly within and between items D and E). The order of the corresponding clauses will, therefore, be related to the serial order of common groups of those clauses.

Because the basic relationships between the parties to engineering contracts are unlikely to vary between one contract and another, the fundamental logic outlined above is unlikely to change, although the logical arrangement of the sections and clauses of the Conditions (within the fundamental logical framework) will vary according to the intentions of the parties and the requirements of the Works.

4.3 Logical Order of Sections

Although engineering contracts cover a wide variety of works, the variation in the forms of the Conditions is relatively small. These generally fall into one of three categories:—

Construction contracts
Manufacturing (or fabrication) and erection contracts
Supply of manufactured items.

The most complex of these categories is the construction contract, involving (particularly in civil engineering) substantial site works combined with either materials supply and specialist sub-contracting or (generally in mechanical or electrical engineering) with design, manufacture and materials supply, often with specialist sub-contracting of manufacture. Some of the sections which go to make up construction contract Conditions may not be required for less complex contracts. However, the sectional framework developed for construction contracts serves to illustrate the development of drafting frameworks for other engineering contracts.

Although many of the clauses needed to deal with the requirements of civil engineering contracts will necessarily differ from those needed to meet the requirements of mechanical and electrical construction contracts, these clauses merely provide the detail for the general terms which are common to all construction contracts. A comparison of standard forms (such as the FIDIC Civil and the FIDIC E & M Conditions) confirms this point. A framework of sectional headings can, therefore, be developed which will be applicable to most engineering construction contracts and which will provide for the major sub-divisions of the terms required to be included in Conditions of Contract.

If the interaction between clauses in different parts of the document is to be minimised, then the arrangement of the sections must comply with the fundamental logic applicable to the document. For this purpose it is obviously desirable for each section to deal only with matters which fall within one item of fundamental logic, but (as pointed out in Part 4.2 above) "the duties and responsibilities of the Contractor" overlap with "the duties and powers of the Employer" so that some sections will deal with matters common to both items D and E. The sections should also, so far as possible, be set out in serial order or in appropriate relationships where no clear serial order exists.

The following table gives a list and order of sectional headings suitable for most construction contracts. It also indicates the items of fundamental logic to which the various sectional headings are related.

Logic Item	Sectional Heading
A	Definitions and Interpretation
B	The Engineer
C	Contract Documents
D,E	Commencement and Programme
	General Obligations
	Labour
D	Workmanship and Materials
	Property in Materials and Plant
	Nominated Subcontracts
E	Completion
	Maintenance and Defects
	Remedies and Powers
	War Clauses
	Contract Price and Liquidated Damages
F	Certificates
	Terms of Payment
G	Settlement of Disputes

This list is based upon the British contract system. It is not intended to cover every possible requirement, nor are all the sections applicable to every type of contract. Thus, in many mechanical and electrical contracts, reference to Prime Cost items is inappropriate, while in international contracts a section will be required to deal with customs duties and import licensing.

Two of the sections which commonly appear in standard forms are not included in this Section list, namely "Assignment and Sub-letting", and "Alterations Additions and Omissions".This is because the logical framework requires these subjects to be dealt with in other sections.

Although Assignment and Sub-letting is, in legal terms, conveniently grouped under one heading, in the practice of engineering contracts there are two separate concepts involved. The intention is that the Contractor who is a party to the contract shall be responsible for the whole of the work and that this responsibility and the consequences that flow from it shall not be passed on to any other organisation, i.e. it is the intention of the contract that the parties be defined in a manner which will ensure that the responsibility for carrying out the work is both legally and factually restricted to the party named as the Contractor. A clause requiring that "The Contractor shall not sub-let the whole of the

Works nor assign the contract without the permission of the Employer" should logically be included with the section defining the parties. On the other hand, sub-letting of parts of the work, particularly to specialist firms, is a normal procedure (subject to certain safeguards) and should require only the consent of the Engineer if he is satisfied that the safeguards have been met; the requirements for safeguards and consent are part of the general obligations of the Contractor and, therefore, fall within that section of the Conditions.

The usual section on "Alterations Additions and Omissions" consists of two parts; the first part is concerned with the power of the Engineer to instruct the Contractor and should be included within logic item B, while the second part is concerned with valuation and should logically be included within item F under the heading of Contract Price.

Similar arguments apply to other matters which appear in some Conditions, such as the proper address and manner of serving notices (which logically falls within Interpretation) and tax matters which may fall within the section for Contract Price or within Terms of Payment.

4.4 Use of Logical Method

The content of each section, comprising the clauses and sub-clauses, depends upon the type of work to which the Conditions apply and the intentions of the party for whom the document is being drafted; such matters as the division of risks between the parties, the responsibility for design, and the insurable risks, will considerably affect the drafting of the document. However, by following the logical arrangement of Sections (which provides for different functions and responsibilities to be separated) each of the requirements and intentions can be analysed to determine the logical items involved and can be drafted accordingly with the minimum of risk of interaction between them. Consequently, while making use of parts of many standard clauses, there should be little difficulty in incorporating special or novel clauses for particular works or particular requirements of the client. An example of client's special requirements would be the Department of Transport's Lane Rental arrangements, which are intended to minimise construction times and provide realistic bonus and penalty provisions. Provided that the draftsman adheres to the principle of limiting the content of each clause to the requirements of the fundamental logic, it is a straightforward exercise to frame appropriate clauses, as follows:

(1) to provide in the Programme section for the Contractor to frame his programme to provide for early completion;

(2) to describe, in the Completion section, which items are to be included in completion of the whole (and, where appropriate, in part completion) and to provide for the determination of the extent of any delays due to causes specifically referred to;

(3) to state, in the section for Contract Price, the payment to be made by the Employer for the work described under Completion, as well as any contra-charges to be made against the Contractor in respect of lane rental or any other payments to be made or charged in respect of bonus or penalty and any additional payments to be made by the Employer for any of the delays due to specified causes; a clear definition of any items of preliminaries which are to form part of payments for delays would also be included;

(4) on the basis of the provisions under Completion and under Contract Price, the requirements for certification of completion and of delays can easily be incorporated in the section for Certificates;

(5) any special financing requirements can be included under Terms of Payment.

One of the important advantages of this method of separating the various requirements into clearly defined sections based upon a fundamental logic is that the draftsman can frame the details of payment to be made with a clear picture of the requirements of previously drafted clauses and can limit payment to the risks that the Employer is prepared to meet. The remaining risks will then devolve upon the Contractor who will have a clear picture of the risks for which he is pricing in his tender. If a standard form of Conditions is written on this basis, it is relatively simple to adjust the payment item to suit the risks the Employer is prepared to take on the particular contract. Such a clear definition of the risks to be taken by the parties should assist in avoiding many of the claims that have become a feature of present day contracting. Of course, the Employer will have to pay for all the risks, either indirectly in the price quoted by the Contractor (who will include in his rates the costs resulting from the risks he has to bear) or directly in respect of those risks borne by the Employer through the contract provisions. However, if the Employer prefers a fixed price, even though the resulting final costs may prove to be higher (should the possible risks not eventuate) then clear statements setting out payments to be made and precluding other payments (for possible risks) should avoid dispute. In general, contracting organisations are practised in taking constructional risks and quantifying them, while many private clients are not able or willing to take such risks, particularly if their interests are mainly in manufacturing. Local authorities are generally better placed than private clients to deal with construction risks, but they are often dismayed to find that claims are made by contractors for matters which they thought were covered in the Contract.

The national standard forms of Conditions do not have any common logical arrangement; some, such as GC/Works/1, do not appear to have any logical arrangement at all, clauses following one another in apparently random fashion. It should not be thought that a lack of logical arrangement of the clauses necessarily denotes an unsatisfactory document. GC/Works/1 and its predecessor CCC/Wks/1 have been in use for more than fifty years and have been revised on a number of occasions so that successive additions have been able to eliminate earlier faults. An engineer who has to draft new or revised Conditions of Contract does not have the advantage of time as a means of eliminating faults and ambiguities and is expected to produce a satisfactory finished document as quickly as possible.

The methods given here (developed from experience of drafting lengthy documents) enable the draftsman to develop a logical framework so that the document can be written in relatively small sections which fit together with a minimum of risk of interaction between the sections as well as an awareness of those inter-relationships which require specially careful treatment to avoid ambiguity. Of course, no method can be a substitute for skilful drafting; any method can only guide and assist the draftsman to arrange and to communicate the requirements for the work in a clear and unambiguous manner.

4.5 Composition of Each Section

The draftsman must be able to recognise the logical section appropriate to a particular clause. Reference has already been made to this aspect of drafting, when attention was drawn to the distinction to be made in an engineering contract between Assignment and Sub-letting. The following examination of the contents appropriate to each section of the Conditions for a construction contract will be of assistance both in drafting construction contracts and in determining the logical framework and contents of other engineering contracts.

4.5.1 Definitions and Interpretation

Although this section comes under the heading of 'logic item A – Definitions of parties to the contract', it is not limited to the definitions of the parties. For practical reasons it is desirable to include within the Section the definitions of those words which have a special meaning within the contract or which are substitutes for repeated phrases. The Section, therefore, comprises two parts:—

(a) A clause defining each of the Parties to the contract, together with a clause or clauses concerning matters of substance which affect the definition.
(b) Clauses concerning the interpretation of the contract, including definitions, interpretive items and administrative matters which are of an interpretive nature.

For the purpose of defining the Employer, Conditions fall into three categories:—

(1) Conditions intended to be used nationally or internationally, which are usually incorporated by reference.
(2) Conditions applying to contracts let by a particular organisation; these incorporate the name of that organisation (as the Employer) in the printed document.
(3) Conditions prepared for particular Works and which, therefore, incorporate the name of the Employer concerned.

Where Conditions are usually incorporated by reference, the clause defining Employer ought to refer to a specific supplementary document in which the name of the Employer is given. For example, the FIDIC Conditions state that the Employer is named in the Part II : GC/Wks/1 has the Employer (defined as the Authority) designated in the Abstract of Particulars. In the ICE Conditions space is allowed in the definition clause for the name of the Employer to be inserted, but this is rarely used because the Conditions are almost always incorporated by reference so that the definition ought to be amended to provide for the name of the Employer to be included in the Special Conditions (Clause 72 onwards) or in some other named document.

It should be sufficient to define the Employer and the Contractor by name, but sometimes a simple definition is added which differentiates between them, such as "The Contractor means the firm or company whose tender has been accepted by the Employer...". The name definition must also include reference to the parties' successors and their permitted assigns. The definitions should not contain any matters of substance concerning assignment (such as reference to consent by the other party to the Contract) or sub-letting (this aspect is discussed in Chapter 3 Part 3.11).

Limitations on assignment should form a separate sub-clause within that part of the section concerned with defining the parties to the contract. A contractor is never permitted to sub-contract the whole of the work in a construction contract; this provision should also be set out in a sub-clause of the clause defining the parties.

The definitions of the parties are fundamental to the contract and the greatest care should be used, when drafting the documents, to ensure that the words "Employer" and "Contractor" are used only in this sense. The usual disclaimer, which says the meanings assigned to the definitions apply "except where the context otherwise requires", should not apply to these fundamental definitions. It is rare to have to refer to an employer or a contractor within the context of the contract and there is, therefore, little difficulty in ensuring that the proper references are made.

The second part of the Section comprises the ordinary definition and interpretation clauses; the definitions set out in Section 61 of the Law of Property Act 1925 (see Chapter 2 Part 2.13, Chapter 3 Part 3.9) should be included for contracts in countries where that or similar Acts do not apply. Although it is common practice for definitions which are applicable only to one section to be included within that section, this can be unsatisfactory because such definitions are often used in other documents. A typical example is the term "Period of Maintenance" (or the "Defects Liability Period") which is often included in the Section concerned with "maintenance and defects" rather than in the interpretation clause. Reference is often required in the Specification and this can be the subject of considerable correspondence during the administration of the contract. For these reasons the definition of Period of Maintenance should preferably be included in the interpretation clause.

As a general principle, all terms which require to be defined in the Conditions of Contract should be included in the interpretation clause except for one-off definitions (which serve a different purpose, as described in Chapter 3 Part 3.12).

The general definitions clause should be followed by other interpretive clauses dealing with such matters as clause references, items to be included within the term "Cost", notices and other similar matters including (where necessary in the UK) application of the Conditions in Scotland.

It is common for lawyers, when drafting documents, to add their marginal notes and headings after the document has been drafted and to provide a clause which states that such headings do not form part of the document. Such a clause is never used in engineering specifications and is not needed where the document is drafted from a logical contents list.

4.5.2 The Engineer

This comes within logic item B and is concerned with the duties of the Engineer and his staff. Although, in most standard forms of Conditions, references to actions by the Engineer and submissions to him are scattered throughout the various clauses and although those Conditions have a section concerning the Engineer and his Representative, there is a lack of a general definition of the duties, responsibilities and powers of the Engineer. The ICE Conditions refers to the Engineer in approximately eighty per cent of its clauses but has no clause dealing with the Engineer as such, only a section dealing with the Engineer's Representative. This is surprising, as the Engineer is a key figure in the English forms of contract. The fundamental requirements that the Engineer should act fairly between the parties and should not be obstructed or prevented from acting fairly, is not dealt with; it is left to vagaries of the common law.

Of all the standard forms, only GC/Works/1 has any summary of the duties of the Engineer (referred to as the SO), although the recently published ICE Conditions of Contract for Minor Works (First edition January 1988) does contain a short list of the Engineer's powers. This summary is limited to the issuing of instructions concerning:

"(a) the variation or modification of the design, quality or quantity of the Works or the addition or omission or substitution of any work;

(b) any discrepancy in or between the Specification and/or Bills of Quantities and/or Drawings;

(c) the removal from the Site of any things for incorporation which are brought thereon by the Contractor and the substitution therefor of any other such things;

(d) the removal and/or re-execution of any work executed by the Contractor;

(e) the order of execution of the Works or any part thereof;

(f) the hours of working and the extent of overtime or nightwork to be adopted;

(g) the suspension of the execution of the Works or any part thereof;

(h) the replacement of any foreman or person below that grade employed in connection with the Contract;

(i) the opening up for inspection of any work covered up;

(j) the amending and making good of any defects in the Maintenence period;

(k) the execution in an emergency of work necessary for security;

(l) the use of materials obtained from excavations on the Site;

(m) any other matter as to which it is necessary or expedient for the SO to issue instructions, directions or explanations."

If, instead of limiting this list to instructions, it is extended to define all the other duties, such as:

(1) watching, inspecting and testing the Works to detect any non-compliance with the Contract;

(2) certifying the Contractor's entitlement under the Contract;

(3) deciding disputes;

then this would become a comprehensive statement of the Engineer's duties. If the Engineer's relationship to the parties is outlined, e.g.:

(i) the Engineer shall act fairly and independently between the interests of the Contractor and the Employer; and

(ii) the parties shall not obstruct or threaten the Engineer in the exercise of his duties of fairness and independence;

then there will be a reasonably complete definition of the functions of the Engineer.

Whether it is appropriate to include a full description of the Engineer's functions will depend upon the nature of the contract, the views of the parties and the opinions of the draftsman concerning the requirements of the work. If, for example, the Employer does not wish to act in accordance with the English contracts system and requires the Engineer to act throughout as the Employer's agent, then this should form part of the Contract; an unqualified tender from the Contractor will confirm his acceptance of this state of affairs. As in all other matters of contract drafting, the wording and arrangement must conform to the requirements of the parties and of the work.

4.5.3 Contract Documents

This section falls within item C of the fundamental logic – Description of Works. The general and detailed descriptions of Works are usually to be found in the Specification, the Bills of Quantities (where included) and the Drawings, so that in practice the Section comprises a list of the Contract documents together with requirements concerning their availability and their interaction.

Many of the standard forms have attached to them a form of Agreement which lists the contract documents; the documents describing the Works are not, therefore, given in the Conditions. However, almost all forms of Tender provide that the tender and its acceptance shall form a contract and it is common, even on very large contracts, for the parties not to sign a formal contract agreement. Only GC/Works/1 sets out the documents which form the Contract, as the first item of the interpretation clause. If there is no formal agreement then a list of the contract documents is essential and ought to form the first clause of this section.

The primary documents are the Tender and the notice of acceptance, followed by the Conditions of Contract, the Specification, the Bills of Quantities (or other payment document) and the Drawings.

It is sometimes suggested that the maximum construction information should be provided on the drawings and that much of the information in the Specification could be more usefully provided on the drawings. This suggestion is based upon the fact that the main documents used in construction are the drawings and that the workmen and foremen are often not fully aware of what is included in the Specification. This is a mistaken view; each document should deal with those matters to which it is best suited. The drawings should deal with the geometrical aspects of the work which cannot be adequately described in words, whereas the specification deals with those particular general requirements which apply to the Contract and with the type and qualities of the materials and workmanship, as well as the constructional and general limitations which are applicable. The drawings are also for displaying information which is best illustrated graphically, such as borehole logs and their location, and the position of survey or setting out points in relation to the Works. Notes on drawings should be limited to instructions and should cross refer to the Specification for particulars of materials and of special requirements.

On large contracts the document draftsman is often not responsible for the drawings, but he should nevertheless check the notes and the subject matter of the drawings to ensure that there are no discrepancies between the drawings and the Specification. If changes are made in the Specification, these may involve changing the clause numbers and it is particularly important not only that cross references in the written documents should be checked but that the drawings should also be checked for the correct cross references to the other documents.

A number of practical matters must be included in this section. They are:—

(i)   The supply and custody of drawings and documents.
(ii) The hierarchy of the documents, i.e. whether certain documents overide others in case of differences between them.
(iii) The means of clarifying any differences between the various documents.
(iv) Providing additions to the documents which may be required to deal with practical problems.

In present day contracts the various documents are usually stated to be mutually explanatory, i.e. one document does not override another, although in certain cases provision is made for the Conditions to take precedence. Either way, a clause should be included to provide that any differences will be explained by the Engineer.

In international contracts it will also be necessary to include clauses dealing with the language of the contract documents and of correspondence, the law under which the contract operates, and the currencies in which payments are to be made.

4.5.4 Commencement and Programme

This section involves both action by the Contractor to speedily carry out all the preliminaries to commencing work on the site (included in logic item D) and the responsibility of the Employer to make available the site, as well as the responsibility of the Engineer to make available documentation and further information which the Contractor requires to commence work (all included in logic item E).

An important aspect of commencement of any contract is the contract planning by the Contractor. This includes both planning the general method of construction, and programming, as well as the provision of plant and the items of work to be sub-contracted. For this reason, programming is coupled with commencement, but completion (which is related to different considerations) is included in a separate and later section.

This section should cover only matters connected with commencement and programme and should be limited to the following:—

(a) The commencement of work and the formal Engineer's order to commence.
(b) The handing over of the Site by the Employer and its taking over by the Contractor.
(c) Submission by the Contractor of his programme and of his method of construction and the Engineer's consent to them.

The provision for submission of a programme is often coupled with a sub-clause dealing with revision of the programme in case of delays during the contract. Revision of the programme is not related to commencement of work; it is concerned with delays which may affect completion and ought, therefore, to be included in the section dealing with Completion, where it can be related more closely to measures which have to be taken to mitigate or avoid delays to completion.

In practice, the clauses dealing with the programme and with method of construction should preferably be simple and general unless the Conditions of Contract are to be applicable only to a limited number of similar contracts. The detailed particulars required in the programme and in the statement of the method of construction are best dealt with in the Specification, which is more closely related to the requirements of the particular Works.

4.5.5 General Obligations

The general obligations referred to are those of the Contractor and this section is not intended to include any of the obligations of the Employer. Its fundamental logic is Item D – Duties and Responsibilities of the Contractor.

The general obligations of the Contractor depend upon the requirements of the Employer and the type of work to be carried out under the contract. However, there are a number of basic items which are common to most construction contracts. Listed (so far as possible) in serial order, they are:—

(a) Sufficiency of tender.
(b) Contractor to provide all things required.
(c) Design by Contractor (where required).
(d) Safety of operations.
(e) Conforming with statutes.
(f) Contractor's superintendence.
(g) Quality of the work.
(h) Sub-contracting.
(i) Setting out.
(j) Information to be provided to the Engineer.
(k) Patent rights and royalties.
(l) Materials arising from site clearance and excavation.
(m) Preventing nuisance and interference with adjoining users.
(n) Transport of exceptional loads.
(o) Care of the Works.
(p) Damage to persons and property.
(q) Insurance.
(r) Facilities for other works.

Among the many other items which are often included in contracts and which fall within this section are — corruption and collusion, confidentiality, site access, pollution, requirements of statutory undertakers, certification of temporary works and progress photographs.

The large number of contractor's obligations arises from the industry's past experience of disputes and from the fact that construction works have to be properly carried out in order to avoid hidden and latent faults which are not easily corrected later. Unlike items of mechanical or electrical equipment, construction work cannot usually be sent back to the manufacturer and changed if it proves unsatisfactory. Most of the items are equally applicable to civil engineering or to mechanical or electrical erection work; many can be combined or replaced by a general clause where this is more appropriate or where simpler forms are preferred. The following explanations of the items will assist with drafting or with modifying the items of existing standard forms.

Item (a) – sufficiency of tender, makes provision for avoiding the type of dispute in which the Contractor says that the Employer knew that certain matters had not been taken into account and that therefore the

Contractor was entitled to extra payment in respect of those matters; it is usual to include a clause declaring that the tender is deemed to have taken account of all matters required in connection with the construction and completion of the Works. Although this clause obviously forms the basis of the contract it does not necessarily have to be included in the Conditions; it can equally well be included in the Form of Tender, with slightly different wording, provided that the Form of Tender is included in the contract Agreement.

Item (b) – Contractor to provide all things required, provides for the second basic requirement which is that the Contractor should provide everything necessary for carrying out the work, i.e. he should not be able to claim that some items are not covered by his tender either because he has overlooked them or because they are not obtainable. This requirement is not vitiated if the contract provides for the Employer to supply some material or plant; the Contractor remains liable in theory if the materials or plant supplied by the Employer prove faulty but will have a claim for such faulty supply which would be equal to the claim by the Employer under the Contract and these claims would, therefore, cancel out so the Contractor would not, in practice, be liable for faults in the Employer's supply items.

Item (c) – design by the Contractor (where required), provides for an explicit statement to be included if the Contractor is to be responsible for design of the whole or part of the Works. This may be included as a sub-clause of the clause which says that the Contractor shall provide all things or (if his design is a major item) in a separate clause. Alternatively, where the Conditions and the Specification have equal status under the Contract, it can be omitted from the Conditions and incorporated in the Specification, but this is not recommended where the Contractor has to design a substantial part of the Works. The Specification may, in any case, have to detail the extent of design to be carried out by the Contractor.

Item (d) – safety of operations, is intended to ensure that the safety of everything connected with carrying out the work is part of the general or basic obligations of the Contractor. Any division of the responsibilities for safety is likely to be counter-productive and the Contractor must be entirely responsible and must be required to set up the necessary organisation to ensure safe working. The safety record of the construction industry, particularly in relation to accidents to workmen at site, is poor; the enforcement of safety provisions which are incorporated in the contract can help to mitigate the problem on individual contracts.

Item (e) – conforming with statutes, is the last of the general or basic obligations and provides for the Contractor to comply with all statutes and statutory regulations and bye-laws. Although the Contractor (like everyone else) has a legal obligation to comply with statutes, the inclusion of a clause in the contract means that a failure to comply is also a breach of contract and that the Employer can claim damages from the Contractor if he is put to expense as a result of the Contract's failure to comply. Many construction operations are the subject of statutory regulations and require consultation with statutory authorities and the payment of fees; the clause concerning compliance with statutes must be drafted so as to draw attention to these requirements and make the Contractor responsible for such consultations and payments. However, it would be vexatious and inappropriate for the Contractor to have to comply with statutory requirements in connection with those things for which the Employer has to be responsible, i.e. for regulations applicable to design not carried out by the Contractor, for the completed Works and in connection with the land on which the Works stands. There must therefore be a proviso removing responsibility from the Contractor for items such as planning permission. Some Employers also require a provision that the Contractor shall comply with common law, but this is unusual.

The other obligations of the Contractor generally stem from the foregoing and may be regarded as amplifications of particular aspects of those obligations. They are equally applicable to all types of contract involving site work and to many manufacturing contracts.

Item (f) – Contractor's superintendence, concerns the quality of the Contractor's site staff and particularly the ability of his agent, which determines whether the work will be done well and profitably. In order to safeguard the Employer it is essential that the Engineer should carefully vet the Agent's qualifications and experience; the clause must provide for this as well as requiring that the Agent shall devote the whole of his time to the work and be empowered to take all instructions on behalf of the Contractor. There should also be provision for a deputy able to take instructions on behalf of the Contractor when the Agent is unavoidably absent from the Site.

Item (g) – quality of the work, deals with the general obligation regarding the quality of work and should be limited to requiring that the work be carried out to the satisfaction of the Engineer. The more detailed requirements should be dealt with in the section dealing with materials and workmanship.

Item (h) – sub-contracting, is intended only to give the Engineer power to refuse to approve a sub-contractor whom he considers not to be acceptable for the proposed work.

Item (i) – setting out, is intended primarily to ensure that the Contractor's responsibility to provide and carry out all things necessary for the construction of the Works includes this important matter. It may be necessary for the Engineer's staff to work closely with the Contractor on setting out and it is important that the Conditions should be drafted so that any liaison or checking by the Engineer or his staff does not absolve the Contractor from his responsibility. The detailed requirement for setting out are best dealt with in the Specification; the Conditions should, therefore, only deal with the general aspects of providing all those things required for setting out and of the Contractor's responsibility for accuracy and for assisting the Engineer in checking.

The foregoing items (a) to (i) represent general obligations which are directly connected with constructing the Works. For large contracts they are usually elaborated further in later sections dealing with Labour, Workmanship and materials, Completion, and Maintenance and Defects. In small contracts, where extensive documentation is not required or might be inappropriate, the general obligations may suffice or they may be subject to some degree of elaboration to cover details appropriate to a particular class of work.

Item (j) – information to be provided to the Engineer, is included to enable the Engineer to monitor that the Contractor is carrying out his obligations. For this, a clause is required in the Conditions dealing with the provision of information from time to time by the Contractor to the Engineer. It will provide the basis for the Engineer to give consent to the Contractor's proposed methods and will generally specify information to be provided concerning:

   (i) revision of the programme for construction of the Works;
  (ii) descriptions of the proposed methods of construction;
 (iii) regular progress reports;
 (iv) details of unexpected physical difficulties;
  (v) particulars of labour, plant, and materials brought on to the Site;
 (vi) advance warnings of commencement of operations both on and off the Site;
(vii) particulars of sub-contracts and of correspondence and discussions with local authorities and statutory bodies;
(viii) documentation of other responsibilities such as insurances.

It is important that the number of copies of documents that the Contractor must provide shall be stated as well as any special conditions concerning particular types of documentation. The draftsman must appreciate that the Contractor can only be required to give information which forms part of his obligations under the contract. This must be spelt out clearly either in the general obligations or in the later sections elaborating those obligations or, in the case of constructional methods, in the Specification. It must also be appreciated that this clause or its later elaboration can include any sort of information connected with the Contract, provided that tenderers are willing to agree to give the information stated. It is possible to include in such clauses that the Contractor shall, upon request, supply the original estimate papers upon which his tender has been based, but it is unlikely that contractors will wish to accept this when tendering. However, many overseas contracts require the tenderer to give a breakdown of all items into labour, materials and plant cost and, if necessary, to demonstrate that the figures given are truly representative.

Most civil engineering contracts provide for additional payments for unforeseen additional difficulties arising during construction. Mechanical and electrical construction contracts usually have similar provisions for unforeseen difficulties in excavation for underground services. In order to be able to check the Contractor's claim concerning such constructional difficulties it is essential, and a prerequisite for the consideration of these claims, that the Engineer is informed as soon as such a problem arises and that full constructional and cost information is made available to him at the time to enable the effects and the cost to be monitored. Entitlement to payment is a separate matter which should be dealt with in the section concerning Contract Price. The draftsman must, therefore, include within the information clauses any requirements for information concerning those constructional difficulties which the Contractor considers are unforeseen and which may give rise to a claim for payment in accordance with the relevant clause in the Contract Price section of the Conditions. The drafting of the clause concerning information can be of considerable financial significance and must be carefully considered in relation to the type of work involved in the contract. Problems have often arisen in connection with Clause 12 of the ICE Conditions (which deals with unforeseen difficulties); in many cases it is desirable to re-think the requirements rather than to slavishly follow drafting of the information required in Clause 12.

Item (k) – patent rights and royalties, and (l) – materials arising from site clearance and excavation, are concerned with safeguarding the parties against claims concerning financial responsibility for materials and operations. The Employer will want to be assured that he will not be liable in respect of claims by allegedly injured parties contending that there has been infringements of patent or copyright in materials and operations used in the Works. The Contractor will be taking up and, in many cases, disposing of materials which are on the site and which, presumably, are the property of the Employer and needs to know to whom the ownership passes in respect of disposal of those materials. The ownership of valuable items such as coins or antiques is usually dealt with in standard conditions, but the ownership of less valuable items such as excavated materials to be disposed off site is often ignored. Where large quantities are involved, the financial implications concerning ownership of materials from excavations or from site clearance may be more important than the few items of high value which may be found. Contracts should clearly state which of the parties becomes the owner of the material which is removed from the site as well as stating when ownership passes from one party to another, e.g. does ownership pass when the material leaves the site or when it is transferred to a third party (such as the owner of a tip) or when it is actually lifted from the site and perhaps transferred to the Contractor's stores on the site? The clauses of item (l) should set out the contract provisions in this respect in order to avoid later disputes concerning items such as minerals or even relatively less expensive items such as top soil. The materials referred to here are not to be confused with materials which are used in the Works and which are dealt with in the section dealing with Workmanship and Materials; the clauses in item (l) should deal only with the Contractor's general obligations to distinguish between those items which continue to be owned by the Employer and those which are transferred to the ownership of the Contractor.

Items (m) – preventing nuisance and interference with adjoining users, and (n) – transport of exceptional loads, deal with environmental problems such as keeping roads clean, tipping only in approved areas, avoidance of pollution (including noise pollution), damage to roads, strengthening of bridges and other structures to carry exceptional constructional loads, and many similar matters. The provisions concerning roads, bridges and structures may be concerned with matters occurring hundreds or even thousands of miles from the site. The clauses should deal with the actions to be taken by the Contractor in connection with these matters, including safeguarding the interests of the Employer. Any payments to be made to the Contractor should be dealt with in the section dealing with Contract Price.

Items (o) – care of the Works, (p) – damage to persons and property, and (q) – insurance, deal with the general action to be taken by the Contractor to avoid damage to persons or property and the safeguards to be provided to the Employer in case such damage takes place. In the case of damage to the Works the Contractor must take prompt action to repair or make good any damage which occurs, in order to ensure completion in accordance with the contract requirements. In the case of damage to property of the Employer or third parties the Contractor should be required to indemnify the Employer against such damage or claims and the clauses should be drafted accordingly. Because claims by third parties may be very large compared with the value of the contract or with the assets of the Contractor, he is normally required to insure against such claims and also to insure against damage to the Works.

Insurance requirements can cause difficult drafting problems and may require expert advice. However, such expert advice must be treated with caution as it is usually related to the practices of the insurance companies with whom individual experts are connected. On many matters there are divergencies of practice between different insurers; insurance clauses must, therefore, be drafted in general terms which avoid being tied to particular practices. Some employers (such as many large oil companies) arrange their own insurances, while many government organisations act as their own insurers. It is essential for the draftsman to obtain full particulars of the Employer's insurance requirements and to ensure that the clauses clearly state which items are to be excluded from the Contract's insurance policies. It is also essential to clarify the requirements concerning exclusions and excesses. Third party policies should normally state the minimum amount of insurance for any one incident and that the number of incidents shall be unlimited. Excesses (i.e. the amount of any claim to be met by the insured) should be subject to a limitation to be stated in the clause. Other matters which require to be dealt with are whether the policy is to be in the joint names of the Contractor and the Employer and to whom the amount of any claim is to be paid. It is also usual to exclude from the Contractor's liability any damage caused by the Employer or the Engineer or their staffs.

Item (r) – facilities for other works, is concerned with the problem, common in construction, of a number of different contractors working on the site at the same time; this is particularly the case with mechanical and electrical erection contracts. In order to avoid disputes it is essential that clauses be included in the Conditions of Contract and in the Specification dealing with this matter. The Conditions of Contract should deal only with the right of the Employer to employ other contractors on the site concurrently and for the Contractor to allow appropriate facilities for such other contractors. The Specification should give detailed particulars of other contracts which will or might be taking place concurrently, with sufficient information for the Contractor to make the necessary provision for co-ordination or for adjustment of his own work.

The foregoing commitments cover the majority of the items usually required under the heading of "general obligations", but there are obviously a large number of other possible items which may be required by the nature of the work or the requirements of particular employers. In the latter case the Employer will often have standard clauses which can be taken over by the draftsman and incorporated in the Conditions. The draftsman has, however, an obligation to ensure that the drafting of any standard clauses supplied by the Employer is consistent with other contract clauses and he may need to modify other clauses or agree modification to the Employer's clauses for this purpose.

4.5.6 Labour

The section dealing with labour is intended to provide safeguards concerning the payment of fair wages, the provision to be made for housing employees, and for the necessary welfare and catering facilities. In the

UK the provisions are usually fairly simple and may be limited to requirements concerning fair wages; until recently this was often by reference to the House of Commons Fair Wages Resolution, but this has now been rescinded and reference is made instead to national agreements between trade unions and employers or (in the case of local authorities) reference to the Local Authority resolution or standing orders. Requirements for housing imported labour (not usually expatriate) and for canteen facilities, vary from contract to contract and, in the UK, are reasonably short and limited; the provisions of labour agreements, of planning controls and of the Health and Safety legislation will usually govern these requirements.

In international contracts there is usually a need for much more extensive provision, particularly in respect of expatriate staff and labour, but this varies from country to country and must be the subject of discussions with the Employer. As an example, the basic provisions in the earlier FIDIC Civil Conditions covered:—

Engagement of labour.
Supply of water.
Alcoholic liquor or drugs.
Arms and ammunition.
Festivals and religious customs.
Epidemics.
Disorderly conduct etc.

There are many other provisions which may be needed, some of which are best dealt with in the Specification. This particularly applies to items such as provision of canteens and of first aid and/or ambulance and hospital facilities, which may require very detailed clauses related to the location of the site.

Although it is common to provide in the Labour section for daily or weekly returns of labour employed, this is best dealt with under the heading of "information to be supplied to Engineer" in the section on General Obligations or in a related section of the Specification.

4.5.7 Workmanship and Materials

This section falls within fundamental logic item D and is mainly concerned with the general provisions for inspection, sampling and testing. The materials referred to should more accurately be described as "things for incorporation in the Permanent Works" and these may range from cement to emergency generators forming part of building services or hydraulic equipment for operating a swing bridge. This precise definition is worthwhile in order to distinguish between such materials and the materials from excavations or demolition or materials which form part of the Temporary Works.

Because of the variety of materials which may be involved, inspection and testing will cover a very wide field, ranging from chemical tests of water to trial erection of steelwork or plant.The detailed requirements can only be dealt with in the Specification for each contract; the Conditions of Contract must be limited to basic generalities such as that the Contractor shall supply samples when instructed and shall provide access for inspection. The subjects to be dealt with are reasonably well defined and may be listed as follows:—

(i) A clause is required stating that work and materials shall be as described in the Contract or as instructed by the Engineer and that the Contractor, shall, if required, demonstrate that they so conform. Because it is not always possible to inspect or test every material or activity (particularly where there is concealment) it is essential to state that failure by the Engineer to disapprove of any workmanship or materials shall not prejudice disapproval at a later date.

(ii) The Contractor must be required to provide samples of constructional materials, i.e. of those "things for incorporation" from which the Works are to be constructed. These samples may be either for testing or for initial approval as examples for comparison with bulk materials which are delivered during the construction of the Works. In the latter case the samples must be supplied packed for storage by the Engineer .

(iii) The Contractor will either have to carry out tests himself or supply the samples to the Engineer or to an approved laboratory for testing. Provision should be made for some tests to be carried out by an independent laboratory whose results will be acceptable to both parties.

(iv) The Contractor must arrange for access to be available to the Engineer and his staff for inspecting all materials at their place of manufacture and, if necessary, to view the manufacturing of the various items. In the case of large items manufactured off the site, the Engineer may require an office to be provided at the manufacturer's works, but this provision is best incorporated in the Specification alongside any other provisions for offices required by the Engineer.

(v) There must be a specific provision entitling the Engineer to inspect any item and this should include the right to require dismantling of manufactured items or uncovering or dismantling of items which have been covered up on the site, as well as trial erection in the manufacturer's works in order to check that the items will be satisfactory when erected on the site.

(vi) A provision is required whereby, if the Contractor cannot promptly remove or replace or make good any item which the Engineer condemns, then the Employer will be entitled to arrange for others to carry out the work.

The power of the Engineer to order additional work is limited to those items which are reasonably required for completion of the Contract. The Engineer would not, for example, have the power to order the construction of a permanent road in a contract for laying cables, unless the contract specifically provided for such a permanent road. However, there are occasions when it is necessary for the Engineer to have exploratory investigations made which are of a character which is different from that of the Permanent Works; exploratory borings to check the properties of the soil, foundations or embankments fall into this category and, unless specific provision is made in the contract for them, it can be argued that they do not form part of the Works and that it is outside the powers of the Engineer to instruct the Contractor to carry out such exploratory borings. Because of this consideration it is usual to provide a clause requiring the Contractor to carry out borings or other exploratory works if so directed; the General Obligations section is the appropriate place for such a clause.

In the case of mechanical and electrical works where tests on completion of the whole or parts of the installation are of great importance, it is desirable to set out in some detail the periods of notice to be given by the Contractor and by the Engineer respectively for testing, for witnessing tests and for related matters.

4.5.8 Property in Materials and Plant

The purpose of this section is to safeguard the Employer against the removal of plant or materials which have been brought to the Site. It therefore involve both items D and E of the fundamental logic. Improper withdrawal of plant or materials from site may give rise to two problems;

(a) the withdrawal may prejudice contract completion; and/or
(b) the plant or material may represent collateral for an advance against "materials on site" and its removal from site would leave the Employer at risk in respect of the advance payment he has made.

It is necessary to include clauses in this section which provide that:

(i) plant and materials brought on to the Site shall be used exclusively for the Works and shall not be removed from Site without the permission of the Engineer;
(ii) all surplus plant and materials shall be removed from the Site on completion of the Works;
(iii) item (i) and (ii) shall not be deemed to give approval of the quality of any work or for use in the Work of any plant or materials brought on to Site.

In the UK, item (i) usually consists of a provision that plant and materials owned by the Contractor and brought on to site by him are deemed to become the property of the Employer but shall revert to the Contractor when the Engineer permits them to be removed from the site. Elsewhere than the UK such a provision may not be legally effective. The provision regarding ownership is not normally included in international contracts.

A related problem arises when advances are made by the Employer against material lying in the Contractor's works or in the works of a manufacturer supplying the items to the Contractor. This situation arises when a substantial part of the Works are goods such as structural steelwork which has to be fabricated off the site or consists of items of specially manufactured plant. The plant or material is then effectively collateral against the advance and, in order to safeguard this collateral, legally effective clauses are required which transfer the ownership of those items of plant or materials to the Employer. Such clauses can be complicated and require legal advice if they have to be drafted specially for a contract; Clause 54 of the ICE. Conditions has been drafted to deal with this problem. An alternative is to require any advance to be covered by a bond from a bank or insurance company, but the bond must be carefully drafted to provide for payment by the bank or insurance company of the amount of the advance "on demand" by the Employer; a bond for "due performance" is unlikely to prove satisfactory as a collateral against an advance payment.

4.5.9 Nominated Sub-contractors

Nominated sub-contractors are peculiar to construction contracts and are not much used elsewhere. The requirements involve responsibilities of both the Contractor and the Employer and therefore involves both items D and E of the fundamental logic.

Sub-contractors may be nominated by name in the contract or may be nominated in the course of the contract by the Engineer in connection with Provisional Sums or for additional work. Where nominated sub-contractors are named in the contract and tendering contractors are able to obtain quotations from them for their work, then no particular difficulty arises unless the sub-contractor fails, in which case the Engineer will have to nominate an alternative. In some cases the Contractor may have to obtain quotations and offer an alternative as part of his responsibilities. Where a sub-contractor is nominated during the course of the contract, complications may arise if the sub-contractor is unwilling to enter into a sub-contract on the terms proposed by the main Contractor and also if there is a failure by the sub-contractor, particularly if the terms of the sub-contract are different from the terms of the main contract. Because of these complications the industry today tries to avoid nominated sub-contracts so far as possible and, as an alternative, to specify either by duty or by product; in the latter case an equivalent product is usually permitted.

The problems with nominated sub-contracts arise from a lack of sufficient information which would enable a tenderer to make firm financial and contractual arrangements prior to award of the contract. The draftsman must overcome these problems by including provisions which are firm where information is available pre-contract, but which are fair to both parties where information is only available post-contract.

If the necessary information is available pre-contract, and it is considered advisable to pre-order, then an order for the work should be placed with the proposed sub-contractor with a provision that the order shall be taken over in its entirety by the main Contractor when appointed. For work in the UK the order should be drafted on the basis of one of the standard forms of sub-contract (because they are known and acceptable to both contractors and sub-contractors) and should include a copy of the appropriate parts of the main contract documents. If the nominated sub-contractor is then named in the tender and a copy of the order is one of the tender documents, the main Contractor will have fully accepted the sub-contractor. A typical clause in the contract would then read –

> The Employer has placed orders for various items of work and materials; these are described in the Contract and copies are annexed to the Contract. The Contractor shall take over each of the orders immediately after the Contract is awarded and, within 21 days of the date of acceptance of the tender, the Contractor shall inform each sub-contractor or supplier that the order has been taken over by him and that he assumes full responsibility as the main contractor or purchaser for the work and/or material. The Contractor shall concurrently inform the sub-contractor or supplier of any adjustments to quantities or to programme which may be required and thereafter the Contractor shall be entirely responsible for the provision of all the work or materials and no further reference shall be made to "orders placed by the Employer". Copies of all correspondence between the Contractor and each sub-contractor or supplier concerning taking over of orders shall be forwarded to the Engineer at the time of despatch or receipt. The foregoing arrangement shall not relieve the Contractor of any of his responsibilities or duties for those items of work or of materials supply in accordance with the Contract.

The sub-contract or supply order can then be treated in every way as though it had been placed by the Contractor on his own initiative, rather than as a nominated sub-contract or supply order.

If nominated sub-contracts or supplies may have to be arranged after the main contract has been placed, then this section should include clauses setting out the procedure to be followed by the main Contractor for obtaining quotations from suitable sub-contractors from a list provided by the Engineer and acceptable to the Contractor. The procedure should set out the requirements of the form of sub-contract necessary to safeguard the interests of the Employer, as well as the procedure for opening the sub-contract offers and for approval and nomination by the Engineer. It is most undesirable for nominated sub-contractors or suppliers to be appointed except as part of the main Contract and on the basis of quotations obtained by the Contractor.

In addition to the foregoing it is essential that the following matters be covered in the drafting of clauses for this section:—

(a) The responsibility for control of the sub-contractor and for instructions to him and for all other matters under the contract shall rest with the Contractor.

(b) The procedure in case of termination of the sub-contract should be set out.

(c) Where a nominated sub-contract includes design, the Contractor shall be responsible for the design of that part of the Work.

In conjunction with (c), it is essential that any nominated sub-contract or provisional item in the Contract which may involve design should have a reference to that design requirement in the Contract documents.

Many standard forms of Conditions include clauses dealing with Provisional Sums in the same section as nominated sub-contracts. Such clauses are, however, concerned with Engineer's instructions for carrying out the work and with payment for it. They should be dealt with either in the section dealing with the powers of the Engineer or under the heading of Contract Price or Terms of Payment.

4.5.10 Completion

Completion involves the responsibility of the Contractor to complete on time and the duty of the Employer (through the certificate of the Engineer) to accept extended times for completion (where provided in the Contract) or to negotiate with the Contractor to extend the Contract to cover acceleration of the work. The section falls within logic items D and E.

The first clause of this Section must require the Contractor to complete the Works and any specified Sections within the times stated within the Contract. The period or periods are usually set out in a separate document as an appendix to the formal Tender or as a schedule attached to the Contract; it is sometimes inserted by the Employer before tenders are called, but may alternatively be inserted by the tenderer as part of his Tender. In many contracts there is provision for a section or sections of the Works to be completed within shorter periods, to enable the Employer to use those sections at an earlier date or to arrange for further work by others to make them ready for use. Sometimes, although there may be no sectional completions stated in the contract, the Employer may take over part of the Works before completion of the remainder; such taking over is usually defined in the Contract as the equivalent of sectional completion.

Completion of the whole of the Works would appear to be fairly obvious and not requiring definition. However, in practice Employers are willing to accept work which has not been completed to the last piece of fence or the last nut and bolt rather than delay taking over. Many contracts, therefore, use the term "substantial completion" or "practical completion" and accept that minor items (to the extent acceptable to the Engineer) may be completed during the Maintenance Period. Such minor items ought to include clearing the Site and it is also desirable that the clauses in this section should give some general definition of the type or extent of such minor items that may be completed later.

In the completion schedule, Sections are referred to by numbers, thus Section 1, Section 2, etc. The detailed definitions of Sections for partial completion are generally so specific that they are best dealt with in the Specification. Those definitions are necessarily related to the completion of particular items, so that the need for "substantial partial completion" does not arise.

There are a number of matters which normally entitle the Contractor to an extension of the completion period. The most important of these arises when a variation order is issued requiring additional work or a change in the method of construction, which causes delay. This section of the Conditions must include clauses which:

(a) set out the matters which give rise to an entitlement to an extension of the completion period, preferably by cross reference to the requirements of the other clauses of the contract; this will require a review of the other contract clauses;

(b) require the Contractor to notify the Engineer when it appears likely that a delay will occur due to a matter for which there is an entitlement to extension of the completion period and to state the effects of that delay;

(c) require the Contractor to minimise the delay where this is possible without additional expense;

(d) require the Engineer to extend the completion period when delay arises due to the matters set out in the clause dealing with extension of time.

Generally, the Engineer is required to extend completion as soon as the probable extent of delay is known. Because of the difficulty of ascertaining the extent of the delay, provision is often made for review of the extension of time by the Engineer, although it is usually accepted that such a review cannot reduce the period of extension because the Contractor will have already arranged his working to suit the extended period previously granted.

Many standard forms of Conditions include within this section provision for liquidated damages if the Contractor fails to complete within the contract period. Liquidated damages logically form part of payment for the work; they are a contra-charge by the Employer for damage he has suffered as a result of the Contractor failing to complete in time and are deducted from payments due from the Employer. Clauses concerning liquidated damages should, therefore, form part of the section concerned with the contract price, which should deal with all payments and contra-charges by the Employer.

4.5.11 Maintenance and Defects

The Contractor is responsible for any defects which arise in the work as a result of defective workmanship and materials. This section of the Conditions deals with that obligation and, in many contracts, is concerned only with those Contractor's duties which fall within logic item D. In mechanical and electrical erection contracts and in landscaping contracts it is necessary for the Contractor to have access to the work during its occupation and use by the Employer; this is for testing and inspection or for cultivation if required. The Employer has a duty to make such access available, so the matters also fall to some degree within logic item E.

It is usual to require the Contractor to make good any defects which appear within one year of completion and this period is referred to as the Maintenance Period or the Defects Liability Period. In the case of landscaping work the Maintenance Period is often two years, in order to ensure that plants and trees are growing healthily before the Contractor is relieved of his responsibilities.

To enable the parties to finalise their responsibilities, the contract normally provides that the Contractor's responsibility for the defective workmanship or materials ends when he has made good any defects which arise during the Maintenance Period. This does not relieve the Contractor of responsibility for latent defects which may appear at a later date or for defects which arise from his failure to carry out the work in accordance with the Contract and of which the Engineer and the Employer were unaware at the end of the Maintenance Period.

In many types of engineering work, the causes of defects which appear at various times may be difficult to ascertain immediately and may require considerable investigation. In those types of work (particularly civil engineering work) it is usual to require the Contractor to repair *all* defects but only to hold him responsible for those defects which are determined by the Engineer to arise from the Contractor's defective work or materials.

In some contracts it may be necessary for the Contractor to carry out normal maintenance as well as repairing defects. It is, therefore, important for the draftsman to define precisely the duties of the Contractor in the Maintenance section.

It may be necessary for part of the work to be removed or for excavations to be carried out in order to find the cause of a defect. The draftsman must include a clause requiring the Contractor to search for the cause of defects, as directed by the Engineer.

In many standard forms a provision is included in the Maintenance section to allow the Employer to arrange for defects to be made good if the Contractor fails to do so within a reasonable time and to charge the cost of such making good to the Contractor. The Employer's remedy in this case could be included within this section, but is more appropriately accommodated in the next section dealing with the Employer's remedies and powers. The Employer's contra-charge for carrying out the work is logically to be included in the section dealing with Contract Price.

4.5.12 Remedies and Powers

This section is primarily concerned with the actions the Employer is entitled to take if the Contractor fails to carry out his responsibilities under the contract and is, therefore, primarily related to logic item E. However, it ought also to deal with the Contractor's remedies against the Employer, such as in connection with the indefinite suspension of the Works. Moreover, as part of the Employer's remedies there may be a duty on the Contractor to take certain actions such as assigning orders for materials to the Employer. There may therefore be a substantial element of logic item D in this section.

The section is one of three in which the draftsman has to review previous clauses to ascertain the matters to be dealt with. In the section on Completion it was necessary to review previous clauses to ascertain the causes of delay as the basis for Extensions of Time; in the section on Contract Price and Liquidated Damages it will be necessary to review the previous clauses to ascertain which are to be subject of a clause concerning payment or contra-charge. In addition, in the section concerning Certificates a review of certain sections will be required in order to define the certificates to be provided by the Engineer and the time at which certification is to take place.

The main matter likely to arise from a review of the clauses of the Conditions will be the Contractor's failure to comply with the Engineer's instructions within a reasonable time, particularly in connection with urgent requirements such as repairs to damaged structures. The usual remedy for this is for the Employer to employ other workmen or contractors; this will involve provision for a contra-charge in the section concerned with payment. Mention was made in the previous section (Maintenance and Defects) concerning failure to remedy defects; this is likely to result from failure by the Contractor to act on an instruction from the Engineer to carry out remedial works.

In addition to the items arising from review of earlier clauses, the Employer needs to be entitled to determine the Contract and take over the Site (if appropriate) in case of bankruptcy or liquidation of the Contractor or if the Contractor delays excessively the carrying out of the Works or on other serious default of the Contractor. The Employer also needs to be able to take over materials supply orders and sub-contracts; the draftsman must frame the clauses so that the Contractor is required to assign the benefits of such orders or sub-contracts to the Employer, should such an eventuality arise.

This section should also include clauses dealing with the Contractor's remedies in case of failure by the Employer to carry out all his duties under the Contract. Most of these remedies are likely to be financial and should be included in the sections dealing with payment, although some contracts give the Contractor the right to determine the Contract if there is excessive delay by the Employer in making interim payments. However, in case of extended suspension of the Works, the Contractor must have the right to determine the whole or part of the Contract, and this right should be set out in suitable clauses as part of Remedies and Powers.

4.5.13 War Clauses

The term "War Clauses" is intended to be a generic term for those situations in which the Contract needs to be determined due to external causes such as war or frustration and also where provision is to be made for unilateral determination by the Employer. It may alternatively be referred to as "Special Powers of Determination".

The Clauses must provide for instructions by the Employer (or by the

Engineer on his behalf) and for action by the Contractor to terminate work both on and off the Site. It, therefore, falls into logic categories D and E.

The section requires the provision of clauses which define the meaning of war for the purpose of the Contract and also the meaning of frustration. The definition of frustration may vary with different systems of law, but in the case of construction it is preferable for the parties to agree that the contract has been frustrated; the definition in the FIDIC E & M Conditions sets this out clearly. It reads –

> '...the term "Frustrated" shall mean the prevention of the fulfilment of the Contract by reason of war or by any cause or causes agreed by both the Employer and the Contractor to be beyond the control of either of them.'

The common requirement of all the matters to be dealt with in this context is to set out measures which have to be taken to protect the Works and the Site and to remove plant and surplus material and leave the Site ready for possible resumption at a later date. This provision is usually in terms of instructions by the Engineer but, in particular contracts, the draftsman may be able to state specific requirements which will automatically take effect. There must, of course, be a provision for the Contractor to carry out all the necessary measures so far as they are practicable in the circumstances.

Other provisions which have to be made are in connection with payments and these should be dealt with in the section dealing with Contract Price.

4.5.14 Contract Price and Liquidated Damages

This section is entirely concerned with payment to be made by the Employer and therefore falls within logic item F. Contra-charges and liquidated damages included within the section represent, in essence, deferred payments or payments made by the Employer to other parties on behalf of the Contractor. This section should be concerned only with entitlement to payment and not with the mechanism of payment or the time at which payment is made, which should be dealt with in the following two sections.

Payments fall under three headings:—

(1) The Tender Price.

    (2) Additional payments or reductions resulting from Engineer's instructions or from actions arising from the operation of particular clauses of the Conditions of Contract.

    (3) Contra-charges which the Employer is entitled to make in accordance with the Contract.

Damages for breach of contract (other than where a specific condition concerning payment or contra-charge is made in particular clauses of the Contract) are excluded from the contract provisions and must be dealt with separately.

The Tender Price is that price which was agreed at the time of letting the Contract. It may be a fixed lump sum or it may be variable, as in a remeasurement contract with Bills of Quantities or based upon a Schedule of Rates. When remeasurement is required to ascertain the sum, clauses must be included in this section setting out the basis of measurement (in the UK this would normally be in the form of a quoted Standard Method) and arrangements for remeasurement. Provision should be made for the Contractor to attend when the Engineer is taking measurements and for preparation of any additional site drawings necessary for measurement purposes.

For those lump sum contracts which do not include a Bill of Quantities or a Schedule of Rates for valuing variations, it is advisable to provide a schedule setting out the price for the various parts of the work, the total of the Schedule being the Tender Price. Such a Schedule will assist in pricing variations and will form a basis for certifying interim payments.

One of the problems in drafting contracts is that, in the course of drafting the substantive requirements, the draftsman becomes so involved in the minutiae of those requirements that it is difficult for him to take a detached view of the financial implications. The effect of excluding financial or price considerations from the substantive clauses and collecting them in a separate section (under the heading of "the Contract Price"), is to allow the draftsman to review the substantive clauses after he has drafted them and thereby to take a more detached view of the financial effects. A review of various draft clauses of the Conditions, to elucidate their financial implications, is in itself a valuable exercise which may result in redrafting of some of the clauses to improve the wording and remove possible ambiguities.

Many of the items of additional payment or savings arise from variation orders issued by the Engineer for additional or reduced volume of work or for varying the type of work to be carried out. It is essential to include clauses setting out the method to be used for valuing variation orders, including provisions for daywork ordered by the Engineer. Valuation of variations in a contract with a Bill of Quantities can be relatively simple, but for lump sum contracts, particularly where these involve design by the Contractor, it may only be possible to define valuation in very general terms and to put the onus on the Engineer to arrive at a fair valuation.

Each additional payment or saving which will be made by the Employer in connection with a substantive clause of the Conditions must be set out in an appropriate clause of this section. Typically, these clauses may include:

(a) disruption due to the failure of the Engineer to give an instruction within a reasonable time, after a request from the Contractor;
(b) additional costs to the Contractor arising from the effects of work by other contractors of the Employer;
(c) additional costs arising due to the action of employees of the Employer;
(d) additional costs arising from insurance risks excluded from the contract clauses;
(e) the costs of uncovering work which proves not to be defective;
(f) failure of the Employer to hand over part of the Site in good time to enable the work to be carried out in accordance with the programme;
(g) any penalty to which the Contractor is subject at law, as a result of carrying out Works in accordance with Contract;
(h) making good of defects for which the Contractor is not liable;
(i) physical conditions which could not reasonably have been foreseen by an experienced contractor.

This list is not intended to be comprehensive; not all of it will be applicable to any particular contract. Applicability will also depend on the Employer's willingness to carry the risks involved or alternatively to require the Contractor to include for those risks in his tender price.

The drafting review and the listing of possible additional payments resulting from the requirements of substantive clauses enables the draftsman to discuss with his client which payments are to be incorporated in the Conditions and which are to be omitted. This procedure also allows the Employer to explicitly decide the risks which he will meet; the Contractor will then have to incorporate the remaining risks in his Tender Price.

It should be noted that some of these matters, such as actions by other of the Employer's contractors, are not risks but are matters for which the Employer is vicariously liable and cannot, therefore, reasonably be excluded from payment.

Special payment provisions are required to deal with the War Clauses section and for any other clauses which provide for determination before completion. Many of the provisions for payment for early determination are alternative to the payment provisions for a completed contract; the draftsman must clearly set out those provisions which may involve alternatives, in a manner which will ensure that they are substituted.

For contracts which are to last more than 2 years it is usual to provide for payment adjustments to take account of market price fluctuations. In the UK there are a number of standard clauses and formulas (Baxter, Osborne, BEAMA etc.), based on published indices, which the draftsman should use. Where there are no suitable published indices the price fluctuation clause will have to be based on market price quotations and the inclusion of a Schedule of Basic Prices of Materials and Labour in the Contract. Clauses will also be required to deal with variations in tax legislation *directly* affecting the contract such as labour-tax matters and VAT. These are best drafted by legal specialists or may be adapted from standard contract forms.

The clauses in this section must set out *all* payments to which the Contractor is to be entitled under the Contract. Many other clauses in the contract may be claimed to give entitlement to some additional payment and, in order to avoid disputes or ambiguities on the subject, it is important to include a clause which states specifically that payment will be limited to the items set out in this section. There must be specific reference to the particular clause numbers providing for payments and to alternative clauses which may provide for payment in respect of early determination. A suitable draft for such a clause might read –

> The Contractor shall be entitled only to such payments set out in Clauses... or in alternative Clauses...and.... This Clause constitutes a waiver by the Contractor of any claim for payment or reimbursement in respect of any provision of any clause of the Contract which may be alleged to be additional to or alternative to the provisions of Clauses.....

The provisions for payment should be followed by clauses dealing with contra-charges. These must relate to specific clauses in the Conditions which entitle the Employer to carry out work which the Contractor fails to do or where there is an obligation on the Contractor which will give rise to additional cost to the Employer if the Contractor fails to carry out that obligation. The number of these items is usually small and comprises:

(a) liquidated damages for failure by the Contractor to complete the whole or a specified part of the Works by the contract completion date or an extended date if an extension of time is granted;

(b) the cost of action by the Employer resulting from failure of the Contractor to carry out an instruction of the Engineer in good time;

(c) the cost to the Employer of remedying defects for which the Contractor is liable and which he has failed to remedy in good time;

(d) the cost to the Employer of bankruptcy or liquidation of the Contractor;

(e) costs arising from a breach of contract by the Contractor which results in determination of the Contract;

(f) the cost of any material, constructional plant or labour supplied to the Contractor by the Employer.

In order to ensure that all these contra-charges are legally deductable it is usual to include a clause or paragraph permitting the Employer to deduct the appropriate sum from any payments due. A typical clause would read –

Whenever under the Contract any sum of money shall be recoverable from or payable by the Contractor, such sum may be deducted from or reduced by the amount of any sum or sums then due or which at any time thereafter may become due to the Contractor under or in respect of the Contract.

This provision is sometimes extended to cover payments due under other contracts with the Employer.

4.5.15 Terms of Payment

This section is concerned with payments made by the Employer and therefore falls within logic item F. The previous section was concerned with the Contractor's entitlement to payment; this section deals with the times at which payments will be made, including provision for advance payments before completion.

The draftsman must bear in mind, when drafting this section, that the

Employer expects to pay the Contract Price when the Contractor has completed the work and made good any defects; the payment of the Contract Price is made when the final account has been agreed. All other payments should be considered as advances by the Employer for which the Contractor must provide collateral in a similar manner to the provision of collateral against a loan. In a construction contract the collateral is usually in the form of:

(1) for an advance payment made before work has commenced or for the purchase of plant or materials (including advances of this type made during the course of the Contract), an unconditional bond or guarantee from a bank or insurance company; or

(2) work and materials which have passed into the ownership of the Employer.

In practice, the only truly unconditional bond or guarantee is one which provides that the bank or insurance company shall repay the amount of the advance payment on the demand of the Employer and is usually referred to as an on-demand bond. This type of bond should not be confused with a bond for due performance which may be provided by the Contractor as a guarantee that he will properly complete the contract. On-demand bonds are usually drafted by the bank or insurance company issuing them, not by the contract draftsman, but the wording has to be checked by the Engineer to ensure that they are truly unconditional.

In many contracts, particularly international contracts for construction, provision is made for an advance payment at the time of signing the contract, this advance being intended to cover such items as the purchase and importation of plant and material. The advance is usually recovered from subsequent interim payments for work carried out and the on-demand bond reduced accordingly, i.e. the collateral against the advance is transferred from the on-demand bond to work and material which has become the property of the Employer. Whenever initial advances of this type are made, the draftsman must provide a detailed recovery clause setting out the period over which the advance will be recovered and the amount be recovered against each subsequent interim payment or by any other system of recovery.

In construction contracts and in other engineering contracts it is usual to make interim payments at regular intervals (commonly monthly) or stage payments when specified stages of the work have been reached. The former method is usually adopted when the work is paid for by detailed measurement; the latter method is commonly applied to lump sum contracts without detailed quantity measurement.

In this section of the Conditions the draftsman must set out full particulars of any measurement requirements for *interim* payments and particulars of when and how these payments will be made. Where only a few stage payments will be made, they may be set out in a clause in this section, but where a substantial number of stages are to be included then it is usually more convenient to provide a separate schedule. A payments schedule is very convenient as an accompaniment to a standard form of Conditions; it allows the draftsman to provide for a wide variety of payments to suit particular circumstances. Where stage payments are in respect of work which is not on the site and which does not become the property of the Employer, then it is common for them to be covered by an on-demand bond. Such bonds can be expensive (the bank may make a charge of the order of between $1\frac{1}{2}\%$ and 2% of the value of the bond). To avoid these costs various legal devices are adopted (instead of on-demand bonds) to ensure that work and materials not on the site become the property of the Employer and cannot be taken over by a liquidator or receiver in bankruptcy. Such legal devices are not practicable in international contracts; even in the UK doubts have been expressed about the applicability of such legal devices in Scotland.

Interim payments against the value of work and materials on the site are not usually made up to the full value of such work and materials, the balance being retained by the Employer as a fund for the remedy of possible defects or for other remedies under the Contract. This retention may amount to between 3% and 10% of the sums concerned and must be clearly set out in this section of the Conditions. Because of cash flow problems affecting contractors, provision is often made giving the Contractor the option of providing on-demand bonds as a substitute for the holding of retention monies.

Because prompt payment is fundamental to the contract, it is essential that all clauses concerning the terms of payment should state when payment is to be made and should provide contractual sanctions if payment is not made in time. These contractual sanctions are usually in the form of interest on late payments but may also include provisions whereby the Contractor may discontinue work if payments are excessively late. Clauses embodying these provisions must be carefully balanced between the desire of the Employer not to be tied down too tightly and the increased cost which will be incorporated in tenders if tendering contractors believe that there are insufficient safeguards in respect of payments.

4.5.16 Certificates

Generally, certificates given by the Engineer are in respect of payments and therefore fall within logic item F. However, certificates are also required in respect of completion and for satisfaction of other requirements which fall within logic items D and E.

Most of the Employer's liabilities are contingent upon the Contractor carrying out his responsibilities. As the Employer is not responsible directly for the administration of the contract (which is carried out through the Engineer) he will, before accepting liability for payment, require a document which will confirm the Contractor's claim to have carried out the contractual responsibilities. This document takes the form of an Engineer's certificate; the clauses of this section must set out all the certificates to be provided and particulars of the matters to be certified in each of them.

As certificates arise from the consequences of other Conditions, the clauses in this section ought to be based upon a review of all the previous clauses of the Conditions. The certificate requirements fall into four categories:—

(a) Payment certificates.
(b) Completion certificates and maintenance certificates.
(c) Certificates for extension of time for completion.
(d) Certificates related to various remedies included in the Conditions.

Clauses concerning payments certificates must define all the items to be included, i.e. whether interim or final payment, any retentions, any other deductions (such as repayment of advances) and any contra-charges. In order to be able to certify the amount due, previous payments must be deducted from total value (less deductions, etc.). As the Engineer is not usually aware of the actual payments made by the Employer, the certificate must be based upon deduction of the total of previous certificates; this should be stated in the appropriate clauses.

Completion and maintenance certificates give the Employer an entitlement to use the Works and, in the latter case, to exclude the Contractor from the land forming the Site. They also usually entitle the Contractor to further payments which have to be the subject of separate certificates to be given within a specified time of the completion or maintenance certificates.

Where the contract makes provision for extension of the construction period if particular specified circumstances arise, the Engineer must provide a certificate setting out the amount of the extension and the reason for it. It is most important that this certificate should give reasons; a particular incident may give rise to a number of claims for extension on different grounds and the parties to the contract require to know which of the grounds for extension comply with the contract and what extension is granted in respect of each. It is usual to include in such certificates the revised date for completion and it is worthwhile drafting this requirement into the appropriate clauses to ensure that the extended period is clearly defined.

Whether any certificates are required in connection with the various remedies which are prescribed, depends on the drafting of the Conditions and must be the subject of a review of the appropriate parts of the Conditions. In many cases the alternative is an instruction from the Engineer to the Contractor, but there may be a requirement for the Engineer to certify if the Contractor has failed to carry out the instructions; the draftsman must check this against the other requirements and draft accordingly.

Certificates for payment and in respect of completion and maintenance always require an application by the Contractor; the clauses concerning the certificates must be drafted accordingly. The clauses for certificates of extension of time should also provide for a claim by the Contractor, but may provide that the Engineer can grant an extension in the absence of a claim by the Contractor or in the absence of specific evidence from the Contractor, in order to safeguard the Employer against circumstances which might make it difficult for him to fulfil his responsibilities if the completion period is not extended.

Certificates in respect of various remedies may not depend on claims by the Contractor. The requirements will have to be ascertained specifically in each case in the course of drafting these clauses.

In many contracts a limitation is placed upon the time available for certification by the Engineer after receipt of a claim from the Contractor, subject to the Contractor providing the whole of the information required for certification. This is a reasonable provision (particularly in the case of payment certificates) to ensure that there is no excessive delay caused by the Engineer. The draftsman must ensure that the Engineer will not be held responsible if the Contractor does not give the necessary information in time.

4.5.17 Settlement of Disputes

Settlement of disputes represents item G of the fundamental logic and is an essential complement to the provisions setting out the responsibilities of the Contractor and the duties of and the payments to be made by the Employer.

Apart from legal action, there are three alternative methods for the settlement of disputes that are available for use on engineering contracts. They are:—

(1) Arbitration.
(2) Expert award.
(3) Conciliation.

Although it is rare for any contract to include all three methods, there is no reason why the section should not provide for each of the three methods to be used for the purposes for which they are best suited.

When drafting the disputes clauses it should be borne in mind that in an engineering contract the parties expect any procedure to provide the best commercial advantage coupled with the lowest cost of the procedure. In many cases, if the procedure is unduly expensive it will not be worthwhile making use of it and the Contractor could be forced to accept an unreasonable judgement by the Engineer or the Employer.

The commonest procedure provided in the standard forms is Arbitration. In UK Conditions it is usual to provide for a single arbitrator agreed by the parties or appointed by the President of an engineering Institution. Elsewhere it is common to provide for a variety of alternatives of two or more arbitrators and/or umpires. The technical and legal requirements for arbitration are dealt with in the clauses of the standard forms; the draftsman should be familiar with those clauses and should ensure that those technical and legal requirements are provided for.

The cost of legal representation at arbitration hearings is often the major part of the cost of arbitration. Verbal evidence and cross-examination are part of the tradition of legal tribunals. The events which give rise to engineering disputes have generally taken place at a much earlier date and oral evidence by the participants is usually based upon their diaries and reports or other documents which they have used to refresh their memories; in the circumstances, oral evidence is often a test of the memory of the witness and of the skill of opposing Counsel in throwing doubt upon the evidence in cross-examination. In many cases, engineering arbitration can be undertaken on the basis of documents only, including the representations of the parties (called "pleadings"), the replies to these representations, and written evidence. Draftsmen should consider including in any arbitration clause a requirement that arbitration should be carried out on the basis of documents unless the arbitrator or the parties require verbal elucidation.

One of the most useful documents when drafting arbitration clauses is the Institution of Civil Engineers Arbitration Procedure, which sets out clearly the necessary steps to be taken and also provides for a short procedure which substantially represents arbitration on the basis of documents and is worthwhile adopting generally. Although the Arbitration Procedure was drafted in connection with civil engineering arbitration, the provisions in it are not specific to civil engineering and are applicable to all branches of engineering.

In the ICE Arbitration Procedure there is a special procedure for experts. In this the experts from both sides prepare reports and then meet the arbitrator and present their case and the questions which they have in connection with other experts' opinion. It is intended for dealing with disputes in which there is a substantial difference of expert opinion on the subjects concerned and where this forms a major part of the dispute. It is an arbitration procedure and is not to be confused with the settlement of disputes by an expert.

There are many disputes where the parties are not divided substantially on matters of principle but cannot agree upon how costs are to be evaluated and upon the total quantum of costs and other items. Such a dispute can be most conveniently settled by the appointment of an expert acceptable to both parties, who will be permitted to examine all the files of the parties and who will determine the cost and other quantity items attributable to each of the parties. Such a determination must be accepted as binding by the parties and must be drafted accordingly. This procedure can also be used for determination of whether the quality of the product meets the requirements of the contract. It can be extended to determine the quantitative aspects of alternatives which may be themselves submitted to arbitration. It is worthwhile including a clause for expert determination of disputes in contracts where the method of costing variations ordered by the engineer cannot be easily defined.

A cheap procedure which is currently being developed, particularly for small contracts, is the use of Conciliation. In this procedure the parties meet a conciliator who may be chosen by them, but is more commonly appointed by a professional or trade organisation. The parties put their cases to the conciliator who may suggest alternative compromise settlements which may be acceptable to both parties. This is followed by a report from the conciliator proposing a specific settlement; it is obviously advantageous for such recommended settlements to have already been accepted by the parties. The ICE has now published 'Conciliation Procedure 1988' which is applicable to most engineering contracts and it is worthwhile providing for the use of the Procedure when drafting a conciliation clause. Such a clause may also provide that arbitration shall only proceed if conciliation fails.

4.6 Other Sections and the Practical Application of the Fundamental Logic

The sectional headings for construction contracts given in the table (in Part 4.3 above) cannot be expected to deal with every aspect which may arise in construction contracts. Additional sections will have to be included to deal with the contract requirements for particular types of work; useful lists of sections which may assist the draftsman are included as Parts 2 of the FIDIC Conditions, but even these cannot be comprehensive. An example of an additional clause specific to a particular type of work which is not listed in the FIDIC Conditions is the provision usually made in highway contracts for the special requirements of Statutory Undertakers. On the other hand, many types of contract may require the omission of some sections described for construction contracts and the substitution of others not listed.

CHAPTER 5

Specifications

5.1 Purpose of Specification

The Conditions of Contract are often referred to as "General Conditions of Contract". This description refers to their applicability to a large number of contracts of the type for which the General Conditions were drafted, rather than to dealing with the general aspects of the Contract. It is rare for Conditions of Contract to be drafted specifically for a particular contract; usually a standard or semi-standard document is modified by amendments and additions drafted to suit the particular needs.

As pointed out in Chapter 4, the Conditions of Contract are concerned primarily with the relations between the parties and only secondarily with technical matters; technical and general matters specific to the contract are dealt with in the Specification. However, there is no fundamental reason why the Conditions of Contract should not also deal with specific general and technical matters: it is merely a matter of convenience that has resulted in this practice. The practice is enhanced by the availability of standard forms of Conditions of Contract applicable to a wide range of engineering work (civil engineering, mechanical and electrical, chemical, etc.) and which provide for a separate Specification to deal with specific general and technical requirements of each contract, thus crystallizing the practice. If one takes the fundamentalist view that the contract consists of a single document applicable to all aspects, the Specification would be written as an extension of or a part of the Conditions of Contract and the whole document merely referred to as the Contract. A compromise on this view was adopted by the chief engineer of a New Town corporation who required specification clauses to be numbered as a continuation of the clause numbering the Conditions of Contract, although the Specification formed a separate document.

In the fundamental logic of the Conditions of Contract (described in Chapter 4, Part 4.2), the Specification forms part of Item C – Description of Works, and would be referred to in this context by cross reference. The other documents which describe the Works are the drawings and the Bills of Quantities (where they form part of the contract). Generally, the purpose of the Specification is to describe those aspects of the Works which can be put into words in a clear and relatively concise manner, whereas the drawings describe the dimensions and form and components of the work and their spatial relationships. Where a Bill of Quantities is included in the contract this merely quantifies the extent of the Works described in the Specification and drawings, in addition to its role in defining Item F – Payments by the Employer, in the fundamental logic of the Conditions of Contract.

Considered in its broadest aspects, the "Description of Works" must include provision for checking that the Works have been properly carried out as well as a means of describing their final extent (after completion and payment), as distinct from the initial extent described in the Specificatiion and the drawings. The Description must, therefore, include every aspect of supervising and administering the carrying out of the Works. In the case of work which is mainly manufacturing, the provisions in the Specification for administering and supervising may be limited to little more than inspection and testing by the Engineer. In the case of contracts involving a large element of site work (such as construction contracts) the Specification may require an extensive general section setting out the information that the Contractor must supply and the provision which must be made for detailed supervision by the Engineer's site staff. It might be considered that many administrative clauses which are included in specifications could form part of the Conditions; with the extensive use of standard forms of Conditions the detailed aspects of administrative clauses may be applicable only to the particular contract and are therefore better dealt with in the Specification.

5.2 Types of Specification

Because of the variety of work which may be the subject of an engineering specification these vary widely in form and content. However, three basic forms can be identified, related to the basic requirements of the contract. These specification types deal respectively with:

(a) the definition of the duty required for the various items in the contract as well as the tests required to confirm that the duty has been met;

(b) design, materials, and workmanship requirements for a contractor to carry out design and construction of the project;

(c) materials, workmanship, and construction or manufacturing requirements where the contractor constructs or manufactures to the Employer's (or his consultant's) design.

Large contracts may require a specification which combines two of the three types, e.g. a design and build specification may incorporate requirements for the duty of many of the items, or a contract for construction to the Employer's design may include a design and build section to be carried out by a nominated sub-contractor.

A duty specification type is usually adopted for the supply of standard or semi-standard manufactured items, often combined with erection in an existing plant or as part of a design and build contract. A typical example would be a contract for a diesel generator for which the Specification would define the duty of the generator and of subsidiary items and would set out the facilities which would be given for erection, as well as the facilities to be provided by the Contractor and the tests which would be carried out to determine that the equipment met the specified duties. The Contractor's tender for such an item would incorporate a detailed description of the plant to be supplied. It is more usual for such items to form part of a larger contract for design and construction of a manufacturing plant, the diesel generator forming part of a sub-contract for supply and erection. Although duty specifications can be relatively simple, the draftsman must ensure that all related matters are dealt with, such as provision for transport, for storage and handling at the site where erection is to take place, for connection to existing utilities and many other points of a similar character which are necessary to ensure that the Works are fully described.

The Specification for a design and build contract is perhaps the most difficult to draft. Unless the requirements and the quality of the various parts of the Works are fully defined, the Contractor is bound to supply and construct the cheapest available. This is unlikely to be advantageous to the Employer in terms of reliability and maintenance. Moreover, there are often a wide variety of alternatives for design and construction

available to the Contractor; the Specification must set out design requirements and codes as well as materials and workmanship requirements
which will cover all these alternatives. In addition, the Specification must
fully define the site and the facilities available there, provide for design
and construction supervision (including detailed checking if required by
the Employer), inspection and testing in any offsite facilities, and final
testing at the site to ensure that the Works are acceptable when
completed. Where the majority of the work is to be carried out in the
Contractors' factory or yard (such as in shipbuilding), design and build
contracts are very common and often require that the Contractor provides extensive facilities on his own premises for supervision by the
Engineer. Where the majority of work is carried out on the Employer's
site, then contracts for construction to the Employer's design are more
common; this is the case in most civil engineering contracts, although
there may also be substantial elements of design by the Contractor.

Design by the Employer's consulting engineer is commonest in civil
engineering and building work, although it is also used in heavy mechanical work where special designs are required for work for which
manufacturers do not have special designs or expertise. Where, as in
civil engineering, the majority of the work is carried out at the site, this
is equivalent to setting up a complete factory for the purpose of carrying
out the Works. The Specification then requires extensive general and
administration clauses to deal with problems arising from this "factory"
on the site, this being the function of the general section of the
Specification. In addition, the engineer (as the designer) is able to specify
those requirements for materials and for workmanship and for factors
governing methods of construction which will enable the design to be
fully realised. Even in this type of contract the Contractor may offer an
alternative design for some parts of the work; in such a case it is common
for the alternative design to be taken over by the Employer or his
consulting engineer who must then add to the documentation, including
specifying any changes in materials and workmanship as well as the
limitations on construction methods needed to ensure satisfactory
completion.

5.3 Logical Arrangement of Specifications

Because the relationships between the parties to an engineering contract
are reasonably similar for a wide variety of contracts, it is possible to

postulate a fundamental logic for the Conditions of Contract which is applicable to most contracts. The types of work which may be the subject of different specifications will vary so widely that it is not possible to postulate a common fundamental logic; the logical arrangement will have to be fitted to the nature and purpose of the particular specification.

Comprehensive coverage of all the detailed requirements of the work is an obvious fundamental requirement for drafting engineering documents. In the drafting of a specification this is achieved by a combination of engineering experience and of a logical application of that experience in drafting.

The engineering experience does not have to be the personal experience of the draftsman. It may be derived from the experiences of one or more colleagues or from the records of organisations responsible for preparing the contract or it may derive from the advice of consultants or from published or other papers on research, construction or manufacturing. For specification writing it is important to be aware of all the practical and theoretical problems which may arise in each item or operation; when the information is derived from sources such as publications or reports, then the draftsman is likely to require wider engineering experience to fill in the obvious and practical gaps which necessarily arise from inability to question the authors.

The purpose of the logic is to enable the draftsman to examine the experience and information in an order which will disclose any gaps and to prepare a drafting arrangement which comprehensively covers all the operations and requirements. The logical method will vary with the type of work concerned and may differ in different parts of the documents. As an example, the most useful logical method in a construction contract would be to examine the requirements of the work in the order of construction operations. In this way, an engineer who is familiar with the construction methods involved will be able to examine each of the construction operations consecutively and specify the particular requirements to suit the site, the design and the contract provisions for the work. This detailed examination by consecutive operations will minimise the risk of omission of an important provision because the specification writer should, by this means, be able to visualise the complete process of construction from beginning to end and thus to identify every operation. In the case of manufacturing or fabricating operations, a similar view based upon an examination of the consecutive operations required in manufacturing or fabricating, will provide the appropriate logic for specifying necessary provisions.

A specification based upon the duties of the various items often presents greater logical difficulties because, although these duties can be identified by consideration of the modes of operation, there are often considerations related to maintenance, life and appearance which are not obvious from a simple examination of modes of operation and which require further consideration in relation to maintenance and damage limitation as well as of appearance and psychological effect on operators.

As an example, the application of the logical method to a specification for a bridge would first identify the order of construction of the various parts as:—

Demolition and site clearance.
Excavation.
Construction of foundations.
Piers and abutments and wing walls.
Filling around foundations, piers, abutments and wing walls.
Bearings.
Deck beams and slabs.
Parapets and other fixtures.
Waterproofing deck and deck finishes.

Each of these parts would then be examined in turn on the basis of the sequence of the operation required for construction. Thus, for example, all the likely methods of excavation would be considered, in the order in which they would be carried out, so as to make provision for:—

(a) Any special requirement to avoid damaging the bearing capacity of the ground.
(b) Special requirements concerning dewatering of the ground.
(c) If any part of the excavation is in a watercourse, requirements for cofferdamming such as depth of sheet piling, minimum height of top of cofferdam, provision in case of flooding, etc.
(d) Examination of completed excavations to confirm that they are satisfactory before further operations are commenced, including provision in case of deterioration.

In the actual drafting of such specifications, many more operations are likely to be considered. These would arise from special problems such as proximity to railways or roads or other structures requiring special precautions. The draftsman will have to decide the particular logical method to be adopted in each case to provide consecutive consideration of operations or items in a manner which will minimise the risk of omission of any likely items or operations.

A specification is not intended to be a list or description of operations or items. In principle, a contractor is entitled to carry out the work in any way which he considers suitable. The specification is intended to provide limitations to the methods, materials and workmanship adopted by the Contractor, either by laying down particular methods or materials or workmanship or by limiting the use of particular methods etc. By a logical examination of each operation or item in turn it is possible to identify limitations necessary to ensure that the design requirements are achieved with reasonable economy and safety.

5.4 Specification Framework

As with other documents, when drafting specifications it is essential to prepare a framework of section headings and clause headings within which the detailed drafting will be carried out. Such a framework should be comprehensive and prepared on the basis of the design and constructional and manufacturing requirements, arranged in a logical order.

The provisions of a specification can be sub-divided into four main groups:—

General requirements, including administrative provisions.
Materials.
Workmanship.
Special requirements for duty or manufacturing or construction.

Many specifications are drafted using these groups as the main section headings, with sub-sections for different materials and for different types of workmanship. In many cases materials and workmanship are not separated, i.e. workmanship applicable to particular materials follows immediately after the requirements for the material concerned. However, the groups listed above represent the types of provision required and do not necessarily, of themselves, form a suitable basis for the drafting framework. Although the "General" group will form a suitable separate section, the other groups merely represent the type of content which must be included within the sectional framework. In addition, where the Contractor is required to provide a substantial design element, it may be necessary to provide a separate section setting out his design responsibilities.

Most organisations preparing engineering contracts have standard specification clauses for materials and workmanship for the classes of work with which they deal, but these have to be modified to suit the requirements of each contract, e.g. if the clauses have been written for use in temperate climates they may require substantial modification for work in tropical climates. Although a specification may contain a section dealing with materials and workmanship, there may nevertheless be a need for further materials and workmanship clauses to be included in the special requirements group. These would deal with work which is applicable to only part of the contract work and are best related to the description of the requirements of that part of the work. For example, there may be clauses concerning the general requirements for fabrication of steel frameworks, but there may also be special requirements for fabrication of a particular part of a structure and this may most conveniently be dealt with in the special requirements for that particular part of the structure.

When preparing the drafting framework, Sections can often be most conveniently arranged according to the operations which have to be carried out in the course of the work, each Section dealing with materials, workmanship and special requirements of a particular operation or group of operations. Another arrangement, where the Specification is mainly concerned with duties of various mechanical or electrical parts, is for each Section to deal with a particular mechanical or electrical item, the various clauses within each section being concerned with the duties of the particular items referred to, as well as setting out the limitations on materials and workmanship needed to ensure that manufactured items are acceptable. Typical single operations which would be the subject of separate sections would be, for example, piling, dredging, steel fabrication, lift supply and erection, switchgear, etc. This arrangement is particularly useful for standard specifications which are intended to be incorporated in the contract with amendments related to the specific contract works.

The art of drafting comprehensive and practical specifications depends to a large extent upon the draftsman's competence in preparing comprehensive and logical frameworks for each contract. Even where the framework of a standard specification is adopted, it is essential that it be modified to provide the framework for the individual contract; standard forms of specification cannot be expected to comprehensively cover all the requirements of any particular contract and a failure to prepare a specific framework in each case may well result in a serious omission.

The following example illustrates the development of the framework of an actual specification which was written for the machinery required for operating a large river control structure. The specification was divided into four sections followed by appendices which consisted of schedules of information and requirements. Three of the sections have been divided into sub-sections to enable requirements for particular parts of the work to be grouped together. The framework adopted for sections and sub-sections was:—

SECTION 1 — GENERAL

SECTION 2 — MATERIALS

    Standards

    Sampling, Testing and Inspection

    Steel and Ironwork

    Paint

SECTION 3 — TRADES

    Standards and Tolerances

    Structural Steelwork

    Welding

    Galvanising

    Protective Systems

SECTION 4 — PARTICULARS OF THE WORKS

    Extent of Works and Contractor's Responsibilities.

    General Description and Requirements.

    Machinery for Large Gates.

    Machinery for Small Gates.

    Hydraulic System.

    Electrical Systems.

APPENDICES

    Special Instructions for Delivery, Erection and Installation of Plant.

    Schedule of Drawings and Information to be Supplied by the Contractor.

    Key Sections and Dates and Other Programme Information.

    Schedule of Tests to be Carried Out at Manufacturers' Premises.

    Schedule of Tests to Be Carried out at Site

    Schedule of Controls, Interlocks, Indications and Alarms.

Each of the sub-sections consisted of a number of clauses detailing the requirements of that sub-section, resulting in a document consisting of approximately three hundred clauses, in addition to the clauses and schedules in the Appendices. It would be impractical to reproduce the full list of clauses and to discuss them here, but the clauses comprising the sub-section for Electrical Systems in Section 4 illustrate the development of sub-sections to form a clause framework as follows:—

Electrical Systems.
4.101 General Description.
4.102 Extent of Works.
4.103 Work Produced by Others.
4.104 Drilling of Steelwork and Fixings.
4.105 Materials and Workmanship.
4.106 Interchangeability.
4.107 Enclosures of Electrical Equipment.
4.108 Protective Finishes.
4.109 System Voltages.
4.110 Electric Motors.
4.111 Fuse Switches and Isolators.
4.112 Contactors.
4.113 Distribution Fuseboards.
4.114 Fuses.
4.115 Cable Types.
4.116 Installation of Cables.
4.117 Cable Glands.
4.118 Cable Trays.
4.119 Conduit and Conduit Fittings.
4.120 Cable Trunking.
4.121 Control Systems.
4.122 Control Panels.
4.123 Pressure Transmitters and Indicators.
4.124 Control Supply Units.
4.125 Gate Position Indicators.
4.126 Limit Switches.
4.127 Relays.
4.128 Control Terminations.
4.129 Earthing.

Although the clause headings of the various sub-sections (other than Electrical Systems) have not been given in the framework above, two clauses in the General section need to be mentioned, namely "Purpose of Works" and "General Description of Works" as well as the first clause in the "Hydraulics System" sub-section which is entitled "Contractor's Responsibilities". In Section 4 – Particulars of the Works, there are sub-sections dealing with "Extent of Works and Contractors Responsibilities" and with "General Description and Requirements". In addition, it will be seen that there are, in the sub-section "Electrical Systems", two clauses under the headings "4.101 General Description" and "4.102 Extent of Works", as well as the clauses referred to in the General section and in the Hydraulic system sub-section. These sub-sections and clauses are not repetitions of one another; they constitute a progression from the general requirements to particular requirements for a limited part of the work. Thus, the clause at the beginning of Section 1 entitled "General Description of Works" gives a general picture of the whole of the Works, but the sub-section in Section 4 entitled "Extent of Works and Contractors Responsibilities" and the sub-section entitled "General Description and Requirements" give specific details of the various items of work and the responsibilities to be undertaken by the Contractor for all items. The description of the Contractor's responsibilities and the requirements given in these sub-sections are broad and are amplified, where required, in particular clauses. In addition, in specialised sub-sections (such as "Hydraulic System" and "Electrical Systems") further detailed descriptions appropriate to those specialities are given.

This arrangement and sub-division is part of the logic of this large specification covering a wide variety of mechanical and electrical work; it assists comprehension by specialists, as well as sub-division of the drafting work so that a team of engineers can carry out the drafting under the supervision of the engineer responsible for the whole of the drafting work. The sub-sections also assist the preparation of sub-contract documents by the Contractor, by segregating specialities which are likely to be the subject of sub-contracts.

A similar logic is apparent in the drafting of clauses dealing with materials and workmanship. The clauses of the sub-section dealing with particular requirements include materials and workmanship which are additional to the clauses in Sections 2 and 3, although, where necessary, they cross refer to those Sections. These materials and workmanship clauses (in particular sub-sections) provide detailed requirements for a limited number of operations, unlike those of Sections 2 and 3 which deal with operations generally required throughout the Works, i.e. in the earlier parts of the Specification the general requirements have been dealt with, while in the latter part of the Specification particular special requirements are given. Thus, in addition to a logic related to order of construction which applies to the clauses in particular sub-sections, there is also a logic running through the whole of the Specification commencing with the general requirements in the early parts and proceeding to the particular requirements in clauses in the latter part of the Specification.

The clauses of the Electrical Systems sub-section have been shown with their original numbers as an illustration of a particular system of numbering. This system, which gives the clauses a prefix number related to the Section, enables the contents list to be sub-divided for ease of reference in a long specification, each of the sectional contents lists being placed at the beginning of the relevant section rather than at the beginning of the Specification. This arrangement is, of course, only useful for large documents where the sub-division assists referencing, particularly in the finding of specialist items.

5.5 Standard Specifications

A knowledge of available standard specification clauses is essential to the specification writer. It allows him to concentrate his efforts on particular and non-standard requirements. At the very least standard clauses act as an aide-memoire, assisting in ensuring the comprehensiveness of the detailed requirements. The uncritical use of standard clauses (particularly by those not familiar with the class of work concerned) can, however, give rise to serious errors concerning the applicability of the requirements to the particular job in hand. Problems also arise from inapplicable references or cross references to other clauses; such references and cross-references must be checked to ensure that they are applicable to the specification which is being written. It is essential that the draftsman be familiar with the clause or have recently read it in detail, so that he can amend any requirements or references. Even those who are continually using particular clauses may often find (generally during proof reading) that they have overlooked some reference or cross-reference when copying a familiar clause.

We may recognise four distinct types of standard specification:—

(a) "Company" standard specifications and clauses.
(b) National standard contract specifications issued by government departments or by national industrial or professional co-ordinating organisations.

(c) Specifications prepared by national standards organisations such as British Standards.

(d) Codes of Practice prepared by national standards organisations or by professional or industrial co-ordinating organisations.

Many large industrial and professional organisations (not necessarily "companies" in the legal sense) have standard specifications for some of the types of work carried out or specified by them. These company specifications are intended to standardise corresponding Sections of contract specifications and are usually appended to the Specifications and introduced by reference. However, in many instances they may need to be amended or overridden by particular requirements and, unless the work is to be absolutely standard, they should be introduced by a clause which states –

Except as otherwise provided in this Specification .... shall be in accordance with Standard Specification Number .... annexed hereto.

Alternatively, where the amendments to the Standards can be succinctly stated, then they may be listed in a clause drafted as follows –

...shall be in accordance with Standard Specification Number... annexed, subject to the following amendments and additions:—

This type of standard specification is common among process contractors, particularly in the oil industry, for such items as steelwork, ladders, stairways, and other items which repeat from plant to plant; it is often accompanied by standard drawings.

A common type of standard used by both large and small organisations, consists of general clauses and clauses for materials and workmanship which are intended to be incorporated in Specifications according to the requirements of the particular work concerned, rather than annexed as a whole. Often the standard consists of clauses carried over from previous specifications (which are treated as standard) rather than being a separate formal standard document. The method adopted in the past was to "cut and paste" the standard clauses which have been selected for incorporation, inserting new material and amendments in the process, to produce a draft Specification. With the advent of the word processor, the same process is carried out by transferring clauses from standard discs or from discs of previous specifications and incorporating new and amended material after the transfer. The word processor has the advantage that a large proportion of the specification will be printed out without further typing, thus avoiding considerable amounts of copy typing and consequent typographical errors. It is, however, important to proof read the *whole* of the typescript (i.e. both the unamended typescript and the amendments and additional matter) to ensure that the clauses read satisfactorily in their amended form and also to weed out or correct any inappropriate references or other unsatisfactory matter which may have been overlooked in the original compilation from the standard clauses. Draftsmen of engineering documents are often under pressure to meet a deadline. If the period available for drafting is very short then there is sometimes a tendency to save time by avoiding proof reading of standard clauses; it cannot be over-emphasised that such procedure is fraught with serious dangers.

National standard contract specifications published by government departments or national industrial or professional organisations are limited in number and are mostly confined to construction work and to civil engineering in particular. Mechanical and electrical work, which deals more with manufactured items, relies for standardisation mainly upon British Standards and on the regulations made by specialist bodies such as the Fire Offices Committee, Lloyds Rules, I.E.E. Regulations, and other similar regulations.

The best known national standard contract specifications are probably the "Specification for Highway Works" (previously the "Specification for Road and Bridge Works") issued by the Department of Transport and the "Civil Engineering specification for the water industry" originally published by the National Water Council and subsequently republished by the successor to that Council. Examples of other national standard contract specifications are the Piling Specifications issued by the Institution of Civil Engineers, and the National Building Specification which is privately published. There are also national specifications used by organisations such as British Telecom. These are not generally published, but copies are available to engineers carrying out work which is subsequently to be used by the organisation although it does not form part of one of their contracts.

The distinguishing feature of all these national standard contract specifications is that they are intended to be used as a whole to form the Specification, subject to amendment and addition and to the automatic omission of clauses which are not applicable. They are also normally incorporated by reference (in a similar manner to standard Conditions of Contract), the actual document not being reproduced in the contract documents. The Department of Transport also provides a standard clause for incorporation of the Specification for Highway Works by reference, which provides for substituted and additional clauses. A similar clause can be used for the incorporation by reference of other national standard specifications.

The national standard contract specifications are drafted to form part of the contract and they, therefore, include the necessary contract references and contractual requirements, including cross references to specific Conditions. If they are to be used with other Conditions then they require amendment of the cross references and of other clauses which may not be compatible with the Conditions of the contract concerned. British Standard Specifications (B.S.'s), on the other hand, do not contain the contractual clauses; each B.S. has a disclaimer which states that it "does not purport to include all the necessary provisions of a contract". Most British Standards deal with requirements for material, processes and/or testing for specific items which may be required for inclusion in various engineering works and are intended to be included in the contract specification by reference to the standard number, date and title.

A typical reference to a British Standard (taken from an existing contract specification) reads "Hydraulic piping shall be Cold Drawn Seamless in accordance with B.S. 778–Steel pipes and steel for hydraulic purposes". The text of this reference gives no date for the British Standard; although dates for Standards are often inserted in the text of specifications, it is common for them to be dealt with either by appending a list of Standards with their dates, at the end of the Specification (as in the Specification for Highway Works),or by inserting a clause in the General section under the general heading of Materials, in the following form –

Except where otherwise specified or authorised by the Engineer all materials and workmanship shall conform to the latest edition of the relevant British Standards Specification.

These alternative methods of dating are used because so many specifications are based upon company or national standard specifications which avoid giving dates of B.S.'s so that provision can be made in contract specification for the B.S. current at the time of the contract. Whichever of the foregoing drafting methods is adopted, the dates of the applicable editions of the British Standards is part of the detailed description of the Works and apply at the date of entering into the contract (in some cases the date of tender may be the date which applies to the edition of British Standards). If any B.S. is revised during the course of the contract, then the Engineer will have to issue a variation order if the work is to be constructed in accordance with the revised B.S. In specifications where reference is made to the "latest edition", the draftsman must check that provisions of such latest editions do not contradict or make a nonsense of the provisions of the specification. An example of this is B.S.1377– Methods of Testing Soils for Civil Engineering Purposes, where the various tests are numbered, but in which the test numbers have been changed in subsequent editions. In consequence, reference to a test by its number in B.S.1377 (a common method in specifications for earthworks) may be incorrect or even nonsensical in respect of a later edition of this Standard.

Many British Standards include alternatives of quality, method, dimension or other properties, which have to be stated in the contract specification; omitting reference to such alternatives may result in serious ambiguity or may even make a nonsense of some of the specification provisions. It is, therefore, essential that the Standards be read and checked, to ensure that the appropriate alternative is included in the contract specification. In some specifications an attempt is made to overcome this problem by inserting a clause on the lines of the following –

> Where any Standard Specification provides for alternatives and no reference is made in this Speciification or on the Drawings to the alternative required by the Works, then the Contractor shall request the Engineer to issue an instruction specifying the applicable alternative. If such instruction is not requested, then any resulting alterations subsequently required by the Engineer to materials and workmanship shall be at the Contractor's expense.

In principle, such a clause introduces ambiguities in the specification, and should only be considered as a backstop to assist in settling omissions in specification drafting rather than to relieve the draftsman of his responsibility to check the Standards which he has specified.

Although in UK contracts references are generally to British Standards, it is not uncommon for other national standards to be used where they are more appropriate or more familiar to the audience responsible for carrying out the work. Thus, for work on process plants it is common to specify A.P.I. (American Petroleum Institute) or A.S.T.M. (American Society for Testing Materials) standards, despite the fact that in many cases British Standards are available for the items. Similarly, on asphalt work many of the standards specified are A.S.T.M., while in some electronic work the appropriate standard may be DIN (Deutsche Industrie Normen). When specifying such standards the draftsman should check that the measurement units used are compatible with other parts of the specification; US standards are generally in modified imperial units while European standards are generally in metric or in SI units. In international contracts, contractors are often permitted to offer alternative national standards, but this requires safeguards to ensure national equivalents; a suitable clause would be in the following form –

Other authoritative standards which ensure an equal or higher quality than the standards referred to in this Specification will be accepted, at the Engineer's discretion. Any such alternative standard proposed by the Contractor shall be submitted in the English language for approval by the Engineer.

British Standard Codes of Practice are generally concerned with rules for design, although they set out some requirements for materials and workmanship. Existing Codes of Practice are being renumbered as British Standards as and when they are revised and new Codes are numbered as B.S.'s. More recently Codes of Practice have been published which deal with matters of safety and pollution control, such as B.S.5228 – Code of Practice for Noise Control on Construction and Demolition Sites and B.S.6164 – Code of Practice for Safety in Tunnelling in the Construction Industry. These safety and pollution codes represent agreed good industrial practice and should always be specified where applicable. However, they do set out alternative requirements applicable to working conditions and it is essential that the specification writer should state the requirements appropriate to the contract works as well as other specific conditions which are to be met. Such additional conditions are almost invariably required; as an example, in connection with the use of B.S. 5228 (which should be specified for all construction or erection works), it is necessary to state the times during which various noise levels are permitted. Where a contract provides for the contractor to both design and construct the Works, then the Codes of Practice for design will form an essential part of that section of the specification dealing with design but will need to be supplemented by rules specific to particular Works.

5.6 Manufacturer's Specifications

Manufacturer's specifications do not normally form part of contract documentation, although they are often referred to in connection with instructions for the erection or use of manufactured items. However, manufacturer's often possess specialist expertise in connection with the products produced by them and, for this reason, parts of the documentation produced by them are often incorporated in contract specification clauses. Moreover, a study of manufacturer's literature will indicate the products which are available, and discussions with manufacturers will indicate the variations on the standard products which can be produced. It is rarely worthwhile specifying a product which is not marketed unless it is considered essential to provide an expensive one-off special for the purpose required.

Manufacturer's specifications generally fall into three categories:—

    (a) Descriptions and properties of the product.
    (b) Recommended clauses for use in contract specifications.
    (c) Instructions for use or erection of the product.

Although category (a) literature is written primarily for sales purposes, it often contains important information, not easily available from other sources, relating to the type of product concerned. The properties given in the literature are often subject to qualifications concerning their applicability and it is, therefore, important to ascertain from the manufacturer the nature of such qualifications before adopting technical data for incorporation in the requirements set out in the Specification. It is obviously preferable to specify a British Standard where this exists for the product but it must be borne in mind that British Standards generally represent the more commonly available items and that many manufacturer's have products whose properties extend beyond the provision of the British Standards or have useful properties not included in the Standard. When it is intended to use the product of a particular manufacturer, then it is preferable to negotiate terms for supply on a subcontract basis prior to letting the main contract and to nominate the product and manufacturer by incorporating terms of the negotiated contract as an appendix to the Specification at the time of tendering for the main contract. Otherwise, where proprietary products are to be incorporated in the Works, the main contractor is given the opportunity of offering alternatives to the product which may be named in the Specification; a typical clause for this purpose reads –

Where proprietary materials are referred to in this Specification by the use of manufacturer's or trade names, such descriptions are intended to establish the type and quality of the materials required. Unless specified by the use of the phrase 'and no other', equivalent materials or equipment of alternative manufacture may be substituted if such equivalent materials or equipment receive the approval of the Engineer.

A practice sometimes adopted where a particular manufacturer's product is desired, consists of specifying all the properties of the product, including certain properties not provided by other manufacturers; such properties may be minor in character and are included to ensure that only one manufacturer is able to supply the items concerned. This practice is to be deprecated as it gives an unfair advantage to the manufacturer concerned and is not in the interests of the Employer.

Manufacturers rarely appreciate the requirements which have to be incorporated in contracts and, in any case, they cannot be expected to know the needs of a particular contract. They usually have sales and marketing organisations intended to maintain and increase the sales of their products and, in this context, they often publish so-called recommended specification clauses, the primary purpose of which is the inclusion of the name of their product in the documentation for a contract. Such recommended specification clauses (category (b)) must be treated with extreme caution; although they often contain useful technical data, it is rarely worthwhile incorporating the clauses as written. Where the particular product is to be specified, the specification writer should draft the necessary clauses independently, although it may be useful to incorporate some of the technical data given in the manufacturer's clauses.

Manufacturers are anxious to ensure that their products are properly used and they normally issue instructions setting out the methods of use which will give satisfactory results. These instructions often include technical information concerning the particular item, but the instructions and information are often not supplied until the goods or the equipment have been purchased. It is usual for the Specification to contain a clause which requires the Contractor to use or install proprietary items in accordance with the manufacturer's instructions. A typical clause would be –

Unless otherwise specified or directed by the Engineer, all proprietary materials, products or equipment incorporated in the Works shall be used and fixed in accordance with the manufacturer's instructions and recommendations.

5.7 General Section

It was mentioned earlier in this chapter that there was no reason in principle why the General section on the specification could not form part of the Conditions of Contract but that, because the clauses were specific to the particular contract, it was more convenient when using standard forms of Conditions to include a General section in the Specification. A large proportion of engineering contracts are either concerned with construction projects or include erection. Where the work is limited to manufacturing or fabrication or where the provision for erection is limited to supplying specialist supervision for erection by another contractor, or where (as in ship building) there are no standard forms of Conditions, it may be worthwhile adding the few general, administrative and inspection clauses to the Conditions of Contract and thus limiting the Specification solely to a detailed description of the work and the technical provisions to be complied with.

The most comprehensive General section is required for those construction contracts which involve extensive site work. Although other types of contract work are likely to require a less extensive General section, the general clauses in them will merely represent a selection of the more extensive coverage required for construction work; the following paragraphs are, therefore, equally applicable to both construction work and to the more limited requirements of other contracts.

The General section of the Specification is primarily concerned with the definition of the Works and of the Site and this must form the basis for the logical framework for drafting. The fundamental logical basis consists of:—

Detailed description of the Works.
Definitions of the Site.
Specific general requirements for design and construction.
Administrative requirements.

For practical drafting some broadening of the logical basis will be required and is likely to comprise the following items:—

(1) Description and extent of Works.

(2) Description of the Site and its contents.
(3) Site facilities to be provided by the Contractor.
(4) Control of pollution and protective measures.
(5A) General requirements for design and checking.
(5B) General requirements for construction and inspection.
(6) Reporting and programming.
(7) Miscellaneous.

In addition, it is often desirable to provide a "Definitions clause" at the commencement of the General section, to define particular terms used throughout the Specification. Where British Standard Glossaries dealing with the subject matter exist, it is worthwhile incorporating their definitions. A suitable clause would read –

> Unless otherwise defined in the Contract the definition of technical terms used in this Specification are those in B.S....– Glossary of ....terms.

Where definitions apply only to one section of the Specification, they may be more conveniently provided at the commencement of that section and a general 'Definitions' clause may not be necessary.

The above items proposed for the logical basis are not intended as section or sub-section headings, but are groupings of clauses set out in an order which will assist the draftsman to ensure comprehensive coverage. The requirements are discussed below.

5.7.1 Description and Extent of Work

All engineering contracts require a detailed definition setting out what is included within the Works. This is most conveniently achieved by the inclusion of a clause entitled "Description and Extent of Works" at the commencement of a specification. This clause will usually comprise three sub-clauses, the first giving a general description of the whole of the Works, the second giving general descriptions of each of the items of work which are included within the Works and the third listing the drawings which are attached to the contract or referring to a schedule where they are listed (usually they are the drawings referred to in the Tender). An example of a clause from one of the contracts for construction of a chemical plant illustrates the type of detail required and the general drafting provisions. The contract is concerned with work in the ground, but piling work is being carried out by another contractor.

Other contractors will fabricate and erect the steel structures and supply the chemical plant items, plant erection being by direct labour. The clause for the ground works reads as follows –

1.0 Description and Extent of Works

1.01 The Works consist of the construction complete of the roads, foundations and drainage for a chemical plant, including the foundations for control rooms and switch rooms, but does not include piling which will be carried out by another contractor.

1.02 The Works generally comprise:

(a) reinforced concrete foundations for the chemical plant and associated equipment;

(b) roads and paving;

(c) foundations and ground floor of control and switchroom buildings, including trenches and trench covers, but not including floor finishes and special covers required for incorporating in such floor finishes;

(d) underground drainage and other pipe systems;

(e) earthworks required for forming bases and bund walls for storage tanks;

(f) earthworks and concrete pipe sleepers required for and in connection with forming grade level pipe tracks;

(g) excavation, backfilling and levelling of trenches for cables and pipes laid or fixed by other contractors;

(h) grouting up under plant and steelwork erected by other contractors and attending upon plant erectors, all as instructed by the Engineer;

(i) all protective and temporary works and other consequential work required for carrying out the foregoing;

(j) such further work as the Engineer may direct for the purpose of construction, completion and maintenence of the Works.

1.03 The Drawings which form part of the Tender are as follows:—
....

In order to ensure that the description of the Works is complete it is necessary to include items of the Temporary Works and other items which do not form part of the Permanent Works as well as any additional work which the Engineer may be entitled to order as a variation order. Items 1.02 (i) and (j) are included for this purpose; similar items are likely to be necessary in most contracts to ensure completeness of the definition.

The Department of Transport, in connection with their Model Form of Contract for Highway Works, omit any clause describing the Works, as it is contended that the Works are described fully by the drawings. In the case of highway works the Department require the drawings to be complete in all details at the time of tendering, but nevertheless there is often insufficient information to provide drawings for some of the items at that time. This problem would in itself justify the inclusion of a Description clause. However, the Department requires that, at the time of tendering, a description of the Works (corresponding to the type of clause outlined above) be included in the Instructions to Tenderers. The Department merely excludes this clause from the formal contract, but does not rigidly apply this to contracts for subsidiary works where a Description clause may be essential. In general, engineering draftsmen are advised to include a "Description and Extent of Works" clause in all contracts except where only manufacturing is involved and the necessary description is therefore covered by the specified duties.

5.7.2 Description of Site

All contracts for work on construction sites or for erection or installation of plant, machinery or services require clauses describing the site of the Works. The Conditions almost invariably define the Site in a manner such that the area of the site is only that area provided by the Employer and does not include working areas provided by the Contractor; the description of the Site in the Specification must not be contrary to the definition in the Conditions. Although a large proportion of the sites for engineering work include working areas for the Contractor, there are many contracts in which no such provision is made and in which any additional area required for offices, stores, workshops or yards has to be provided by the Contractor. This applies particularly to highway contracts where the Site is usually wholly occupied by the Works and the Contractor will usually have to rent or acquire land locally to provide for his working areas. Any special requirements concerning such additional working area arranged by a contractor must be specifically written into the Specification, but right of access for the Engineer does not usually require specific reference as it will ordinarily be covered in the Conditions.

The site description will usually comprise three clauses as follows:—

Location and Extent of Site.
Site Data.
Site Access and Use.

The clause headed Location and Extent of Site will usually consist of two or more sub-clauses, the first dealing with location and stating the position of the Site in terms which will enable it to be identified on a map when read in conjunction with the site location shown on the Drawings, thus –

The Site lies in the county of Somerset approximately ...km. east of ...... on the ...... Road.

The second sub-clause defines the extent of the site and may be limited to a statement such as –

The Site comprises the area shown hatched on the Site Plan on Drawing Number ...

or, in more complicated cases, may consist of a list of numbered or lettered items preceded by the phrase –

The Site comprises:—.

The Site may consist of a number of areas, some of which can be shown on the drawings and others which may need to be described because the precise boundary will have to be set out on site jointly by the Engineer and the Contractor. On some areas the period of usage may be limited; this must be clearly stated to avoid disputes. There may be other areas which have to be made available, for items such as connection of services, where the precise location cannot be defined until existing underground services are uncovered. In other cases, such as work in a river, the boundary of the site within the river may be subject to agreement with the Authority controlling the river, who may not be willing to finalise the matter until full particulars, including the Contractor's working methods and plant, are available; in such a case the area will have to be defined as being within the boundary to be agreed with that Authority. Where the contract is for installation of machinery or services, the Site boundaries may have to be defined in terms of maintaining a minimum clearance from existing machinery or other installations.

All these points and others which may define the site area must be set out in detail so that the Contractor is fully aware of the area available to him for carrying out the work and can make arrangements to work within that area or to provide other working areas if he considers this desirable and possible. It will also enable the Engineer to enforce any limitations of the period of availability or of the area of the various parts of the Site. Where, within the various items of the sub-clause defining the Site, the information or description needs to be extensive, the draftsman may find it more convenient to merely list the items and add further sub-clauses defining them or providing additional information concerning them; such an arrangement will improve the readability of the document.

Conditions of Contract usually provide that the Contractor shall be deemed to have inspected the Site and made himself fully aware of all matters connected with it which may affect the carrying out of the work. In theory, therefore, the Contractor should be given no information concerning the Site other than the location and extent and should be expected to make all necessary enquiries before entering into a contract. In practice, the Employer is likely to have a considerable amount of information (concerning the Site) which the Contractor requires for carrying out the work, and the contract will need to be based upon this information. If the information is not made available the result may be uneconomical working and higher cost or claims by the Contractor that the matters concerned could not reasonably have been foreseen and that the Employer should be liable for the additional cost concerned.

The best professional practice requires that all information or physical data concerning the site which may affect the carrying out of the work should be made available to the Contractor and should be included within the contract in order to provide formal assurance that the Contractor is responsible for taking account of this information. It is most important to make available the whole of the factual information or data in the possession of the Employer; partial information may prove misleading in certain circumstances (such circumstances often cannot be foreseen) and may give rise to a claim of misrepresentation. It is not necessary to provide information which is publicly available, although it may be worthwhile drawing attention to such availability or to require the Contractor to make enquiries with a particular public authority. Interpretation of data should not be provided; where such interpretation cannot be separated from factual data, then attention should be drawn to the interpretation and it should be excluded from the contract. The Contractor must interpret the data in relation to his requirements and should not be able to claim that he had been misled by interpretations made by the Employer or the Employer's advisors. The information or data may be given on the drawings and referred to in sub-clauses of the clause 'Site Data' or may be provided in other parts of the Specification and cross-referred to in 'Site Data' or, alternatively, the sub-clauses of that clause may state where and when the data is available for inspection.

Where the site work to be carried out by the Contractor is limited to installation of machinery, plant or services, site data may consist only of descriptions of the site conditions under which work will be carried out and the clause will be more appropriately titled "Description of Site" or "Working Conditions at Site". In the case of installation or erection work it is essential to state all matters which will affect erecting, i.e. whether the working floor will be concrete or other hard surface, whether scaffolding will be available, whether cranage will be available and if so under what circumstances and at what times and any limitations of use, etc. The draftsman should bear in mind that the state of the structure when the installation is carried out may be different from that at the time of tendering by the Contractor; the clause must obviously set out the specific conditions which will arise during installation.

On contracts involving substantial site works, the requirements concerning access to the Site and the Contractor's use of the Site will entail sufficient drafting to justify an additional clause entitled 'Site Access and Use'. Where requirements are simple, perhaps amounting only to "Access to the site shall be only through the gate on ... Street", such requirement can be incorporated as a sub-clause in the Site Data Clause or (where the requirements are so limited as not to justify a Site Data clause), incorporated in the clause 'Location and Extent of Site'. Requirements and limitations on access and use can vary widely with different types of work and different sites. The subject requires careful analysis to ensure a balance between economy and necessary limitations to safeguard the parties to the contract. Some of the important questions which the draftsman must consider are:—

(a) Should there be limitations on the Contractor's routes to the Site in order to avoid undesirable disturbance.

(b) In consequence, should there be limitations on the number and/ or position of access points.

(c) What provision is to be made for car parks for Contractor's employees and should it be a condition that they shall not park on roads surrounding the Site.

(d) Where the Site is in more than one area (such as in an existing factory where the office and storage may be separate from the erection areas) are there limitations on the routes between the areas.

(e) Will the Contractor have to share his access routes across the Site or between sites with other contractors or other parties and will the Contractor be required to provide additional access routes within the Site for other contractors.

(f) Are there any Local Planning Authority requirements in respect of access to and on the Site which have to be met by the Contractor.

(g) Are there any bridges or other structures (such as suspended floors) which the Contractor may have to strengthen.

(h) Should there be any limitation on the weight of the Contractor's plant or vehicles to safeguard underground services (particularly in factory areas) or limitation on plant height to avoid possible damage to suspended services.

(i) Will the Contractor be required to provide a temporary hoarding or to fence around the Site and/or around working areas or provide temporary barriers around areas where plant is being erected and, if so, what height and type of hoarding or fencing or barrier is required.

(j) On a green fields site, should the Contractor be legally responsible for the control of "injurious weeds".

(k) Should the Contractor's attention be drawn to his responsibility for the issue of notices for particular matters concerning access, such as "Notices to mariners" in the case of a river or coastal works.

The foregoing are merely representative of matters concerning access to the Site, each of which might require the drafting of a sub-clause. Many of them may not be applicable and there are likely to be many others peculiar to particular contracts and sites which will have to be included. There are also a number of matters concerning the use of the Site which should be dealt with, namely:—

(l) If there are to be limitations on access routes to the Site then the Contractor should be required to make this a condition of all his sub-contracts and of all his orders for supply of material and constructional plant; if this is not included there may be considerable difficulty in enforcing requirements concerning limitation of routes.

(m) As a legal safeguard, the Engineer and the Employer should be aware of any private arrangement for access to the Site, made by the Contractor with adjoining landowners; a sub-clause should be included which requires the Contractor to make such arrangements in writing and to forward a copy immediately to the Engineer.

(n) There have been cases where contractors have used the Site for carrying out work in connection with contracts for other employers in the area, particularly for the prefabrication of temporary works; to safeguard against this practice, a sub-clause should provide that the Site shall be used only for the purpose of the Contract.

(o) The Contractor should be made aware of the probable timing and nature of other contracts or work on or adjoining the Site during the currency of the Contract, particularly where there is likely to be interaction between various contractors and more particularly where one contractor has to provide facilities for another; in complex works involving a number of mechanical, electrical and civil engineering contracts this matter is likely to require a further clause rather than a sub-clause to "Site Access and Use".

(p) Where public or private services cross the Site, the Contractor must allow access to them by the parties owning the services, at any time.

There may be a need for additional clauses if some of the provisions which have to be included would involve lengthy sub-clauses and sub-sub-clauses. If a clause has too many or too lengthy sub-clauses it becomes tedious to read and difficult to find any particular provision quickly; by breaking a clause into two, three or even four clauses with appropriate sub-clauses the general drafting will be improved. An alternative might be to incorporate some of the provisions in schedules or appendices, but this should preferably be limited to matters which can be dealt with separately within the Contractor's organisation, or by sub-contract.

Most of the provisions under the headings of "Site Access and Use" are fundamental to construction and cannot be conveniently sub-divided. However, the facilities to be provided by the Contractor can often be dealt with in schedules or appendices, as discussed below.

5.7.3 Site Facilities

Generally in construction contracts the Site facilities for use by the Contractor's staff and labour and by the Engineer's staff (and often by other contractors also) are to be provided by the Contractor. In certain cases some of these facilities are provided by the Employer, but most prefer all costs connected with the Works to be included within the cost of the various contracts.

Usually the provision of site facilities can be dealt with in three or four clauses, as follows:—

(i) Contractor's Working Areas and Facilities.
(ii) Engineer's Offices and Facilities
(iii) Engineer's Off-site Facilities
(iv) Facilities for Other Contractors.

Clause (iv) will only be required where there are a number of interlocking contracts involving work by different contractors.

Although the extent of the Contractor's working areas within the Site can be dealt with in the clause which sets out the extent of the site (and even where the working area is detached from the remainder it may be more convenient to include it in that clause), it is usually more satisfactory to draft a separate clause which combines both the definition of the working areas and the requirements regarding the facilities which the Contractor must provide for his own work. The clause 'Contractor's Work Areas and Facilities' should have sub-clauses defining the working areas available to the Contractor on the Site together with any requirements for temporary fencing or hoardings or barriers needed to safeguard the Employer. In the UK today there are statutory requirements dealing with safety, health and welfare on construction sites, but many of these are in general terms. Public authorities and large corporations often have a policy of including more specific requirements in their contracts in order to safeguard the public and persons having business on their sites, as well as to ensure that welfare facilities are adequate. In the category of protection of persons, the most common requirements are:

(a) that the Contractor shall provide safety helmets complying with BS5240 and shall make it a condition of employment that all his employees shall wear them on the Site;
(b) site electricity supplies and installations shall comply with BS4363;
(c) where explosives may be required, provisions for storage and use;

(d) provisions for the storage and use of flammable gases;

(e) where necessary, provisions for the storage and use of toxic substances.

On the subject of welfare provisions, the more usual sub-clauses deal with:

(f) specific requirements for the provision of toilet and washing accommodation;

(g) on large contracts, specific provisions for first aid room and ambulance arrangements;

(h) on large sites, canteen and other arrangements for the supply of food.

The specifying of canteens is a debatable subject; site canteens often prove unsatisfactory and give rise to complaints and action by contractors' employees which may be detrimental to the work. Any sub-clause on the subject should be flexible and should avoid giving detailed requirements for the administration and furnishing of the canteen.

Unless there are existing buildings available on site or, in the case of very large works, the Employer decides to construct a separate building under a preliminary contract, accommodation for the Engineer is invariably provided by the Contractor. Depending upon the size of the contract, the Contractor may be required to provide facilities, generally as follows:—

(a) One or more fixed office buildings, each with toilet accommodation; on extensive sites, additional portable offices with toilet accommodation.

(b) On large or highly technical contracts, a laboratory which may require a separate building with special storage accommodation.

(c) Where the contract requires the storage of large quantities of samples (such as drilling cores and samples), separate properly equipped storage buildings.

(d) Furniture for use in all buildings.

(e) Equipment for use by the Engineer (this may include laboratory equipment, surveying equipment, and instruments and testing equipment), but on small contracts it may be more appropriate for the Engineer to use the Contractor's equipment.

(f) Cleaning and maintenance of all buildings and the provision of toilet supplies.

(g) The supply and maintenance of vehicles.

(h) The supply of protective clothing for use by the Engineer and his staff.

(i) On extensive sites, the provision and maintenance of radio communication equipment.

(j) The provision of workmen to assist the Engineer when required.

(k) The supply of specified documents such as British Standards.

On small contracts, where the site supervisory staff may be limited to one resident engineer, the provision of a simple hut, its furnishing and equipment can be dealt with in a relatively short clause with a few subclauses setting out detailed particulars. To avoid later dispute, all services and every item of furniture and equipment must be stated. In large contracts this will involve extensive lists of items which are best dealt with in appendices or schedules, the drafting of the main clause being limited to the general requirements and being cross referred to the appendices or schedules. A typical clause for a large contract would include sub-clauses dealing with the following matters:—

(i) The Contractor is to provide, maintain and remove the offices, laboratories, and stores and their contents, access roads and hard standings as well as equipment, vehicles and documents all set out in the schedule to the clause.

(ii) The offices etc. to be provided and removed at the time stated (in the sub-clause) or when directed by the Engineer.

(iii) The position in which the offices etc. are to be erected (usually adjacent to the Contractor's area) is to be subject to approval of the Engineer. If they cannot be placed on the Site, then the sub-clause must say who has to provide the land required for the Engineer's office compound.

(iv) The offices etc. to be regularly and properly cleaned daily and provided with adequate supplies of toilet materials.

(v) The types of workmen or operatives required and the general requirements for their work such as overtime when required, labourers carrying messages, operatives who require driving licences, etc.

In large contracts the schedule to the clause is likely to need sub-divisions because of its length, the sections of the schedule often being titled as follows:—

A. *Engineer's Representative's Offices, Laboratories and Stores* – this would include an outline specification for the buildings and their services, including layout sketches giving the minimum areas of the various rooms.

B. *Furniture for Offices and Laboratories* – this would include a general specification for the types of furniture and a list setting out the number and description of each item required. The items do not have to be related to parts of the buildings; their use can be decided when the furniture is to be delivered.

C. *Equipment to be Used Solely by the Engineer's Staff* – this may require further sub-division such as surveying equipment, laboratory equipment, electrical instruments, mechanical and electrical equipment.

D. *Protective Clothing* – the numbers of various items should be sufficient to allow for a reasonable number of visitors in addition to the site staff.

E. *Communications Equipment for Use Solely by the Engineer's Staff* – the schedule must state, for each item of equipment, who is responsible for obtaining licences for use, for payment of rental and/or hire charges, and for payment of call charges.

F. *Vehicles for the Engineer* – this could include road vehicles, boats, and in some cases helicopters or aeroplanes. The numbers and types will need to be listed and it should be stated who has responsibility for insurance (business use by any driver approved by the Engineer, for motor vehicles), fuel, lubricants, maintenance and replacement of faulty vehicles.

G. *Documents* – generally British Standards and any other standard document referred to in the contract.

On a recent large contract the schedule for the Engineer's facilities was twenty five pages long. It was, therefore, essential to sub-divide what amounted to a small separate specification. It would have detracted from the readability of the main specification clause if it had been included in the body of the text instead of as an appended schedule.

Where a large proportion of the work included in the contract is to be carried out off site either in a manufacturer's works or in a yard or a dry dock, there may be a requirement for the Contractor to provide facilities there for the Engineer's supervisory and inspection staff. For work likely to be carried out in an existing factory or yard, the requirements for facilities will have to be stated in a fairly general form, as it is unlikely that the particular factory or yard will have been selected at the time of writing the Specification; the requirements will often be limited to the provision of a small office and the use of available facilities. Such requirements can usually be drafted as a sub-clause within the clause "Engineer's Office Facilities". Where the office work may be carried out in temporary yards, the requirements for Engineer's staff facilities will be similar but additional to those required at Site and are likely to require an additional clause for "Engineer's Offsite Facilities". Such a clause could cross refer to the Site Facilities clause and, if necessary, cross refer to parts of the Schedule to that clause for specifications of buildings, furniture, etc. Because the Specification is usually written before a contractor is appointed, there may be considerable uncertainty as to whether a large part of the work will be carried out offsite or on the Site and the specification writer will have to consider all aspects of that problem and make provision for the possible alternative contingencies in the Offsite Facilities clause. As an example, in a particular contract it appeared likely that certain very large floating items might be manufactured in dry docks overseas and towed into position for installation on site. The Offsite Facilities clause had to take account of this possibility and of the problem of supervising a long distance tow in open sea conditions; in the event the items were built in a specially constructed dry dock at the Site and floated out and as a result that Offsite Facility clause was largely redundant.

Where the work includes substantial amounts of civil, mechanical and electrical engineering, the various types of engineering work are often let as separate contracts and provision has to be made for one contractor to provide facilities for others, in order to avoid a proliferation of small safety and welfare provisions; this is particularly the case with toilet, first aid and access facilities. These may be provided by the Employer and constructed under a preliminary contract, but it is often difficult to plan and position such facilities without a knowledge of the various contractors' working methods and labour forces. An alternative strategy is for one contractor (usually civil engineering, as the first large contractor on the site) to construct the facilities he requires and to allow other contractors to use them and also to extend them at the request and the expense of other contractors. This arrangement will require clauses in the specification for the other contracts, referring to the availability of the facility and the provisions for extending the facility, as well as for limiting the entitlement of the other contractor to construct additional facilities. This limitation can be enforced by making such additional facilities subject to approval by the Engineer and requiring them to be available for use by all contractors.

In the case of first aid room and ambulance facilities, adequate facilities sufficient for all contractors may be specified to be provided by one contractor (in positions to be agreed on site) when there is likely to be a sufficiently large total labour force.

In practice, on large and complex works, a combination of these methods of dealing with common facilities is adopted; a comprehensive clause setting out all the requirements may be needed. There may also be a need to provide for the Engineer to arbitrate between contractors in case of dispute concerning the extent and cost of the facilities they require from one another.

5.7.4 Control of Pollution and Protective Measures

The control of pollution and the avoidance of nuisance to premises adjoining the site is a sensitive matter today and any misdeeds by a contractor could result in action against the Employer. Prevention is very much better than cure and, in contracts where there is a possibility of pollution or nuisance it is, therefore, appropriate to specify preventive measures and procedures. Protection of existing service mains and structures (particularly those belonging to statutory authorities or to neighbouring establishments) is logically related to measures for the avoidance of nuisance and should be included within this part of the Specification. Some of these measures may also be applicable to protection of the Works which should, therefore, also be included in this part of the Specification.

The necessary provisions can usually be covered in four main clauses:—

Avoidance of Pollution and Nuisance.
Construction Noise.
Protection of Existing Services and Structures.
Protection of the Permanent Works.

There are a number of statutory requirements governing the control of various forms of pollution; it is not necessary to include in specifications general statements prohibiting a contractor from polluting water courses etc. However, the statutory requirements are mostly in very general terms and it is often desirable to impose some detailed restrictions or to require detailed discussion with statutory authorities so as to ensure that the proper measures to avoid pollution are taken by the Contractor. Discussions may have taken place with water authorities or other local authorities in the course of preparing the contract documents; any measures agreed with them must, of course, be incorporated in the clause dealing with "Avoidance of Pollution". An example of the type of measure which might have to be specified would be the provision by the Contractor of settling ponds for all water discharging to a stream or to sewers and drains.

The Clean Air Act, where it has been adopted, provides some control of atmospheric discharges, but does not control the discharge of toxic materials. Should there be a risk of such a discharge, either in the course of construction at the Site or accidentally (as, for example, during the delivery of materials to the site) then the Contractor should be required to submit proposals for temporary works or procedures which will prevent such discharges and will deal with them if they are accidentally released despite precautions.

The measures to be taken to prevent possible pollution may be complex and politically sensitive; if in doubt, it is always worthwhile specifying that the Contractor shall discuss necessary measures with the appropriate authorities (who should be specifically named in each case) and should either do this in conjunction with the Engineer's Representative or should inform him immediately of the results of such discussions and act promptly on any measure agreed with or required by the appropriate authority.

The avoidance of nuisance to the public and to adjoining owners is closely related to the avoidance of pollution. The three most common causes of complaint and the provision for dealing with them are:

(i) deposition of mud and droppings on the road from vehicles delivering to or leaving the site – a sub-clause requiring the Contractor to provide and use apparatus for cleaning lorry wheels, bodywork and roads should be included where there is a risk of this type of nuisance;

(ii) tipping of site spoil on unauthorised sites – this is a serious problem on many city sites due to the activities of some unscrupulous transport sub-contractors; on sites where this possibility arises, a sub-clause should be included requiring the Contractor to prevent spoil or rubbish being dumped on other than a "recognised tipping area" and making the Contractor responsible for stating where material has been dumped (when so requested by the Engineer) and for removing any material improperly dumped and transporting it to a recognised tipping area;

(iii) dust from the site settling on to adjoining property – where dusty processes may be required in the construction or erection of the Works, a sub-clause should be included requiring the Contractor to provide and use plant for the extraction of dust from the appropriate processes.

Another related matter is the protection of services and structures on the Site. Once again, prevention is very much better than cure (and cheaper to both parties). Specific measures should be required of the Contractor where a possible risk of damage is anticipated; such measures would be set out in sub-clauses to a clause entitled "Protection of Existing Services and Structures" and should include:

(a) provisions for the support of underground services (to the satisfaction of the owner of the services) during excavations;
(b) protection of underground services from the effect of heavy loads arising from the movement of plant and/or from the storage of materials;
(c) protection of buildings, of structures and services above ground, of items of archaeological interest, of permanent markers, etc., by boarding them up or otherwise to the approval of the Engineer;
(d) protection of trees and other valuable natural items by fencing off or other similar means.

The clause may need to include a general requirement for protection of existing structures by methods approved by the Engineer, in addition to specific requirement against identifiable risks.

Although the emanation of excessive noise from a site would in the past be classified as a nuisance, it is today classed as "noise pollution" and is subject, in the UK, to statutory requirements. The detailed limitations under the Act are laid down locally by the environmental health officers of the local council. It is advisable to consult these officers when framing a specification clause which provides for a contractor to take measures to ensure that acceptable noise levels at various times of the day are not exceeded and that there is regular monitoring to confirm that the specified requirements are being adhered to. Complaints and protests by local residents concerning excessive noise levels can have a serious effect on the progress of the work and may be damaging to the Employer's interests. Clauses dealing with "Construction Noise" in inhabited areas have become relatively standard; a typical clause would read –

"1. Without prejudice to the Contractor's obligations under the Contract, the Contractor shall comply with the general recommendations set out in B.S.5228 – Code of Practice for Noise Control on Construction and Open Sites and shall comply in particular with the requirements set out in the Schedule to this Clause.

2. The Contractor shall furnish such information as may be requested by the Engineer in relation to noise levels emitted by constructional plant installed on the Site, or which the Contractor intends to install on the Site, and shall afford all reasonable facilities to enable the Engineer to carry out such investigations as he may consider necessary in connection with noise emissions on and from the Site.

3. If, at any time, the local authority requires compliance with construction noise conditions which are more onerous than those set out in Schedule to this Clause then such compliance shall be entirely at the Contractor's expense."

The Schedule to the clause would usually set out particular requirements on the following matters:—

(a) The provision of exhaust silencers, sound insulated enclosures, special "sound reduced" plant and provision for shut down or throttling down of plant in intermittent use.

(b) Limitations on the times of use of particularly noisy plant (such as pile driving plant) which is difficult to effectively silence.

(c) Specific noise levels at various times of the day and of the week which should not be exceeded on average (12 hour Leq.) and at peak at the site boundary, or alternatively at one metre from the facade of adjoining buildings, together with the requirement that the Contractor shall minimise noise emission so far as reasonably practicable and particularly from operations permitted at night.

(d) Requirements for monitoring noise emissions, setting out the frequency of monitoring by the Contractor and the submission of reports on monitoring to the Engineer, together with the method of monitoring (such as that set out in the Appendix to B.S.5228 Part 1, on a sound level meter to B.S.4197 set on slow response).

If the noise levels specified in the Schedule comply with the local environmental health officer's requirements, then the local authority is unlikely to set more onerous levels unless there are complaints resulting from unsatisfactory or insensitive control and liaison by the Contractor. Making the contractor responsible for the expense of complying with more onerous levels is a recognition of his responsibility in this matter and should encourage liaison with and sensitivity to the noise problems of local residents.

Many site construction processes may cause damage to the Permanent Works which is difficult to remove and therefore merit specific protection measures. Typical examples of such damage are rust staining on visible concrete facing or the entry of abrasive dust into sensitive mechanical parts. Where such risks arise it is advisable to provide a clause requiring the Contractor to provide protection and specifying the type of protection to be provided, as well as requiring the submission of particulars to the Engineer for his consent.

The measures which the Contractor must take to avoid pollution and to protect existing structures and the Works depend upon his working methods and risks. The purpose of specifying is to avoid the possibility of the Employer becoming involved in these matters, which are so dependent upon the Contractor's actions. The clauses concerned must, therefore, be drafted in a manner which will avoid reducing the Contractor's responsibility and risk. If there has not been careful attention to this aspect of drafting then the Contractor may be able to contend that he has a claim against the Employer in case of any action against him by an aggrieved Authority or local resident. The appropriate wording is indicated in the above typical clause concerning noise. As this part of the Specification is concerned primarily with safeguarding the Employer, it is important that the specification writer consult his client on the extent and type of measures to be specified. Generally, most public authorities will wish to have specific measures included in the document, but some private clients may prefer to leave these matters to be covered by the general clauses of the Conditions of Contract and to face any consequences that may arise later, in the belief that this might prove more economical. If the measures specified have been carefully thought out and reasonably related to the risks, then there ought not to be any resulting increases in cost.

5.7.5A General Requirements for Design and Checking

A large part of the General section will have to deal with the general requirements for construction, fabrication and erection of the Works.

In the case of a contract for design as well as construction, that part of the General section will have to be preceeded by clauses laying down the general requirements for design which the Contractor must adhere to.

The Contractor may either be required to design all aspects of the Works (as in so-called "Turnkey" contracts) in accordance with specified duties or may be provided with an outline design in the form of layout drawings and a general specification and be required to carry out the whole of the detailed design including, if necessary, adjustments to the general design if these are required in order to provide satisfactory details. One example of a turnkey contract is for a process plant where the duties consist of the definition of products and by-products to be produced as well as effluent quality (from a specified range of feedstocks), the Contractor being free to produce his design subject to construction or quality standards laid down in the Specification.

An example of a different turnkey basis is represented by a contract for design and construction of a multi-story car park in which the specified duty is the number of cars to be accommodated and the Contractor is free to design the car park subject to the layout, traffic and safety criteria laid down in a design manual (issued by the local Authority) and subject to specified requirements for type and quality of building services and of structural design standards.

An example of a design and construct contract where the Contractor was limited to the detailed design was the case of a pier to be constructed with shops and amusements, in which general design layout and elevation drawings (prepared by the Employer's architect) formed part of the Contract and the Contractor had to prepare his design to provide the completed pier in conformity with the overall design layout and the architectural, structural and services criteria laid down in the Specification. There was no duty specified for the whole of the Works, but there were duties specified for particular parts, such as number of persons entering (for entrance design), provisions for security, service connections to shop and amusement premises, etc.

The overall extent of design requirements should be stated as an item in the sub-clauses of "Description and Extent of Works", which sets out the items comprised in the Works. The item dealing with design would be near the end of the list and would refer to those previous items for which the Contractor is responsible for the design. Where the Contractor is responsible for designing all items comprised in the work, then a typical design item in the sub-clause would read –

The detailed design for the whole of the foregoing including the preparation of drawings and their submission together with calculations, specifications and other documents, for approval by statutory and other authorities and the Engineer.

Detailed design criteria for particular classes of work will have to be given in each of the technical sections of the Specification. In the General section, the specification writer will have to provide clauses setting out the criteria for the whole of the Works and the requirements for submission and approval of the design. Three clauses are likely to be required, entitled respectively –

Design of Permanent Works.
Form of Design Documents.
Approval of Design Documents.

In the case of turnkey contracts, the design criteria for the whole of the Works should be set out under the heading of "Design of Permanent Works" and reference made to design criteria for parts of the Works (such as services) specified in other clauses. Where only detailed design is required in accordance with a general design shown on Tender drawings, the clause will have to state that –

The Contractor shall carry out the whole of detailed design of the Works in accordance with the general design outlines shown on the Drawings referred to in the Tender and in accordance with the criteria and the requirements set out in the Specification.

The design criteria will have to be elaborated to make it clear that the Contractor's design work must include all documents required to obtain statutory and Engineer's approval and to satisfy all necessary criteria whether stated explicitly in the contract documents or not. The Contractor should also be required to carry out any further investigations needed to determine additional criteria (such as soil data) to enable the design to be satisfactorily completed. Alternatively, if the Employer prefers to carry out such investigations himself, then there should be provision for the Contractor to notify the Employer of the investigation requirements in good time to avoid delay. The clause should, if possible, specify any design life and any related maintenance intentions, e.g. whether design life is to be based on minimum maintenance or on maintenance at specified periods. The clause will also have to state any general requirements concerning standards and codes of practice to be used in design.

The Employer will normally expect the Engineer to check the Contractor's design, but the extent of such checks will vary from contract to contract. The clauses concerning the Form of Design Documents and their approval should be drafted so as to ensure that the appropriate documents (drawings, calculations, diagrams, computer printouts, and specifications) are provided by the Contractor for all the checking required and in sufficient numbers and appropriate form to enable the Engineer to check and comment on them and to file and store them efficiently. The sub-clauses will usually specify the maximum size of drawings, the standard titles and information to be provided on the drawings, the size and titling of calculation sheets and specifications, etc. The sub-clauses will also have to state the parts of the Works for which the Contractor must submit drawings and other documents for approval and the procedures for amending the design and for the issue of drawings "Approved for Construction". Because of the close connection between design, fabrication or manufacture and construction, if a fault arises in the Works it will be difficult to identify the precise cause and it is therefore essential that in the drafting of design clauses the Contractor shall not be relieved of any of his responsibilities for design, fabrication or manufacture and construction. A typical sub-clause would be –

The Engineer's approval or consent to any drawing or any other document shall not relieve the Contractor of any of his responsibilities under the Contract nor shall it be interpreted as confirming or otherwise the accuracy of any dimension or level shown on such drawings. In particular, and without prejudice to the generality of the foregoing, the Engineer's approval shall not be deemed to override or replace any approval by a Statutory or other Authority which may be required.

The Engineer's approval merely confirms that an approved item will be acceptable if, when incorporated in the Works, it complies with all the requirements of the Contract.

Where there is no contractor's design element in the contract there may also be a requirement for the submission, for approval, of working drawings of fabrication and of manufacturer's drawings. A clause will therefore be required setting out the numbers and types of drawings to be submitted and the procedure for approval; the clause will be similar to that required for a design and construct contract and should include the provision that approval will not relieve the Contractor of any of his responsibilities.

It is common during fabrication, manufacture and construction for detailed amendments to be made to the Works and many Employers require "as-built" drawings to be provided by the Contractor on completion of the Contract. For this purpose a clause will be required setting out the numbers and types of drawings and other documents to be provided; these are often microfilmed and it is most convenient for the Contractor to arrange this, when the "as-built" drawings have been completed.

Whenever drawings or other documents are to be provided under the contract, there should be provision for contents lists to be supplied for easy reference.

Where the Contractor has to provide electrical or mechanical plant, the Employer will require operating and maintenance manuals. An additional clause will be needed setting out the general requirement for these manuals, such as number of copies, size (usually A4), types of covers and bindings, titling, indexing, etc. Any technical requirements to be specified for inclusion are best dealt with in the clauses which specify the plant duties and requirements rather than in the General section.

5.7.5B General Requirements for Construction and Inspection

Whether or not the Contractor has any responsibility for design, he will always have responsibility for manufacture, fabrication and/or construction and the general requirements and limitations on these matters will have to be described in the General section of the Specification. The clauses will usually fall into four main categories: —

Materials and Quality Assurance.
Temporary Works.
Setting Out.
Construction Methods and Good Housekeeping at Site.

Most of the clauses apply to erection and construction at the site; general clauses giving requirements and limitations on manufacture and fabrication may be required in the General section, but are often more conveniently placed in sections dealing more specifically with particular aspects of manufacture and fabrication.

Where the Specification has a separate section concerned solely with materials, it will be more appropriate for the general clauses concerning materials to be incorporated at the commencement of that section;

otherwise, it is convenient for the clause to be incorporated in the General section. The general requirements for materials will usually cover five subjects:—

(1) Standards to be complied with – this may consist of a general statement requiring British Standards to be complied with unless otherwise stated or permitted, or alternatively it may consist of a list of specific standards applicable to the work which must be generally complied with unless otherwise specified or instructed, the latter alternative often being favoured in mechanical and electrical contracts. There may also be a requirement that only new materials shall be used in the Permanent Works.

(2) Sampling and testing – dealing with:
  (i) supply and delivery by the Contractor of samples for testing and of representative samples for approval;
  (ii) responsibility for payment for samples;
  (iii) acceptability requirements for testing laboratories, testing of samples and supply of test certificates;
  (iv) responsibility for payment for testing and/or test certificates;
  (v) arrangements for plant or machinery testing at manufacturer's works, for acceptance testing and for final tests of the Works.

(3) Factory and other off site inspection of materials, of fabrications, and of machinery and plant.

(4) Use of proprietary materials (including requirements to follow manufacturer's instructions) and also the use of alternatives to those specified.

(5) Transportation and storage on site or intermediately – this subject may overlap with provisions in the Conditions of Contract (such as Conditions concerning exceptional loads) and with specific requirements in the technical sections of the Specification; it is often limited to dealing with the subject of special packing, of storage off-site, of covered storage and of keeping materials clean on the Site.

Quality Assurance (QA) is a subject of increasing importance for manufactured goods and is now being extended to construction work at site. Manufacturer's QA schemes can be the subject of approval by the British Standards Insitution to confirm that they comply with the appropriate QA standards. The adoption of any approved QA scheme reduces the need for extensive inspection at the factory and it is, therefore, in the Employer's interest that the Engineer should be aware of all those items which are subject to such QA schemes. A suitable clause dealing with this matter would read –

> The Contractor shall obtain from manufacturers of equipment which is proposed for incorporation in the Works, particulars of the quality control systems used in the manufacturer's works, including particulars of any national quality assurance schemes with which the manufacture is registered and of the operation of such quality assurance schemes.

This clause makes provision for imported items as well as items which may be the subject of "British Standard" or other schemes.

There are often special requirements for the Contractor's Temporary Works, such as minimum loads and widths of bridging to allow access by the Employer, certificates for temporary structures, cofferdam design standards, etc. and these may require one or more clauses in the General section.

Clauses dealing with setting out of the Works are usually applicable primarily to civil engineering and especially to highway works. There may, however, be requirements in connection with the erection of mechanical and electrical plant, including providing for holding down bolts and fixings to be set out by the erection contractor so that they can be installed accurately by the civil engineering contractor. The clauses concerning setting out should also deal with the provision of instruments to the Engineer for checking (where there is no other provision for such instruments elsewhere), availability of temporary marks, pegs, etc. used by the Contractor in setting out, and particulars of setting out stations and bench marks provided by the Employer together with any connected system for transferring these to the Works. There will also need to be requirements for maintenance of setting out stations and bench marks throughout the period of the contract and for the construction of any permanent marks which may be required.

There are a number of construction operations and related matters which are not conveniently included in other general items or in the technical sections of specifications; these can be grouped together as "Construction Methods and Good Housekeeping". They will not generally apply to manufacturing or fabrication but would be required in connection with site erection. Because of the variety of works involved, there will be a wide variety of miscellaneous items and it is possible to refer to only a few basic points which are likely to be applicable to most construction contracts, namely:

(a) keeping the Works dry – clauses dealing with excavation usually provide for draining or pumping out water; however there are other construction operations which will be affected by water and a general clause is, therefore, usually required;

(b) control and clearing of rubbish at site – a proliferation of rubbish on the Site, coupled with poor layout of plant and operations, gives rise to dangers and inefficient working; the Engineer should have power to lay down areas for the storage of rubbish pending removal from the Site;

(c) pre-construction surveys – some construction operations may give rise to claims for damage by the public or by the Employer and it may be necessary for joint surveys to be undertaken prior to commencement of construction, for comparison in case of damage claims;

(d) notification of construction operations – to enable the Engineer to arrange inspection of construction operations it is essential that these be notified by the Contractor before work on them is commenced; a similar provision is required for manufacturing and fabrication operations. There must also be provision for copies of material orders and sub-contracts to be made available to the Engineer, when required, so that before inspection visits are undertaken it can be confirmed that the orders placed by the Contractor comply with the requirements of the Specification.

The basis for item (a) can be a simple clause such as –

The Contractor shall not allow water to accumulate in any part of the Works or to affect any construction operation; water arising from or draining into the Works shall be drained or pumped to an approved disposal point.

This simple clause can be elaborated to suit any specific requirements in connection with the whole of the Works; it may not cover all the requirements for specific operations, such as excavation, and additional clauses may be necessary in the excavation section of the Specification.

5.7.6 Reporting and Programming

This part of the General section is concerned with the Engineer's specific administrative requirements for the Contract. The requirements for notifying construction operations could be included in this part but are usually more conveniently placed in clauses dealing with construction requirements. Generally two clauses are required, entitled respectively –

Reports, Returns and Records; and
Programme.

The clause concerning Reports, etc. should have an initial sub-clause setting out the number of copies of all reports, returns and other statements and drawings which the Specification requires the Contractor to submit. Usually two or three copies are required (one may have to be returned to the Contractor with comments) but in some cases additional copies are required for distribution in the Engineer's organisation and to clients and other outside organisations who must be kept informed. Subsequent sub-clauses should set out the various general returns which have to be made by the Contractor such as daily returns of labour, staff and plant, weekly and monthly progress reports, meteorological reports, etc. Detailed technical reports or records such as geological strata found in excavations, reports and records of material and plant testing, and the like, should be dealt with in the technical sections of the Specification, where full details of the content required in such reports can be given. All sub-clauses dealing with reports and records must state the times within which the reports and records are to be made available.

The Conditions of Contract will usually have clauses which requires the Contractor to submit to the Engineer, within a specified time, a programme for carrying out the work and provision for early completion of defined Sections of the Works. The Programme clause in the Specification must set out the particular requirements of the contract programme as well as the details to be included within it; it is an essential clause in all engineering specifications. As in the case of many other general clauses, it could be added as an additional clause to the Conditions but, because it is so specific, it is more conveniently included in the Specification General section. The format of the Programme clause can be standardised as six or seven sub-clauses, as follows:—

(1) Stating the form in which the programme is to be submitted i.e. bar chart or time location chart or critical path diagram or other form or a combination of these, together with times for submission. Supplementary documents, such as schedules of drawings, may be required for design and construct contracts. As an example, the following has been used on a number of contracts –

"The programme required by Clause... of the Conditions of Contract shall be in the form of a bar chart covering all the main items of design and of construction and testing of the Works at the Site, laid out in a format which will permit progress of the various items to be indicated on it throughout the execution of the Works. It shall be supported by a critical path network of all requisite operations, which shall be submitted with the programme. The programme shall also be accompanied by a schedule of drawings which the Contractor expects to prepare, giving the number and title of each drawing, its expected date of submission to the Engineer and the date by which approval will be required."

This typical clause can easily be modified to suit the particular requirements of each contract; the references to design and to drawings prepared by the Contractor can be omitted where the Contractor is not responsible for design.

(2) Giving the specific dates by which items of information are required from the Contractor and from the Engineer. The following is a typical example of a sub-clause which was used for a particular contract –

The programme shall show, inter alia:

(a) the dates by which drawings for the various parts of the Works are to be submitted to the Engineer for approval and the dates by which approvals by the Engineer and by the Statutory and other Authorities are expected;

(b) the dates by which the Contractor requires further information for design and construction, in accordance with the provisions of the Contract;

(c) the dates by which the Contractor requires instructions from the Engineer to carry out work described in the Contract as Provisional Sums;

(d) the dates by which the Contractor will require to place any nominated sub-contracts;

(e) the periods for design and the construction of important Temporary Works.

This sub-clause is more specifically related to the contract details than the first sub-clause and has to be carefully drafted to suit the requirements of each particular contract, e.g. in the case of an overseas contract, dates for import of equipment and materials may be required or there may be a need to plan dates by which approval is to be given for setting up supplementary working areas off the site, in order to obtain permission from the relevant government departments.

(3) Listing particular requirements which the Contractor must take into account when preparing the programme for the Works; these requirements are very specific to the particular contract and vary considerably from contract to contract; the specification writer must carefully analyse all matters which may affect the timing of the various operations and draw attention to those which may affect the Contractor's working arrangements. Depending on the circumstances, the length of the sub-clause may vary from a few lines to several pages, particularly where there are interactions between adjoining contracts, or where railway possessions are required, or where operations involve public safety and which may require special working conditions. The following example was used in the Specification for a design and build car park –

The Contractor shall take account of the following requirements in the preparation of the programme:

(a) the construction of the Permanent Works shall not commence until the Contractor has received approval, from the Engineer and the Statutory Authorities, of the calculations and drawings for the layout, building foundations, details of basement floors, walls and waterproof membranes, and details of smoke ventilation and floor drainage to basement floors;

(b) service connections involving excavations in public highways may require 24 hour working;

(c) landscape work will have to be carried out at a time to enable young plants to become well established; if necessary, provision shall be made for temporary irrigation pending completion of the permanent irrigation system.

(4) A method statement will normally be required and this sub-clause should draw attention to matters which are to be included within the method statement; many standard forms of Conditions have a clause providing for submission of a method statement and the sub-clause should, in that case, include a reference to the clause number, as in the following typical sub-clause –

The methods statement required in accordance with Clause...

of the Conditions of Contract shall set out the manner in which the Contractor intends to carry out the construction of the Works, an outline of the temporary works to be provided and a list of the major items of constructional plant and their use in the Works.

Other items, such as particulars of off-site yards, arrangements for the supply of particular types of materials, arrangements for co-ordination with equipment suppliers, etc. may require to be included in order to ensure that the method statement gives a clear picture to the Engineer to assist him in arranging for the appropriate supervision.

(5) There are often differences between the Contractor and the Engineer as to whether the programme is sufficiently comprehensive and takes full account of all the likely problems; a sub-clause is required to enable the Engineer to insist upon additional information if he considers it necessary, on the lines of the following –

The Contractor shall supply such supplementary information requested by the Engineer as will enable the Engineer to satisfy himself that the programme complies with the requirements of the Contract and may consequently be approved. If the Engineer is not so satisfied the Contractor shall, after receipt of the Engineer's observations, modify the programme in a manner which will so satisfy the Engineer and thereafter the Engineer will approve the programme. The same procedure shall be carried out in connection with any revised programme submitted by the Contractor.

(6) The Contractor will usually have to submit, for the monthly progress meetings, a progress chart comprising a copy of the programme marked up with the progress of each operation. A typical sub-clause would read –

The Contractor shall submit to the Engineer monthly (one week before the Monthly Progress Meeting) three copies of the approved programme with the progress of all items of work at the date of submission marked thereon and shall concurrently submit two copies of the supporting critical path network with progress marked up similarly and with such modifications as may be required to ensure that subsequent operations are completed in accordance with programme requirements.

(7) Many forms of Conditions have clauses which require the Contractor to inform the Engineer from time to time as information is required for constructional purposes. The programme clause should avoid prejudicing this provision; a suitable sub-clause would read –

> The provision of the information above shall not absolve the Contractor from obligations under Clause...of the Conditions of Contract.

5.7.7 Miscellaneous General Clauses

Most contracts have a number of general requirements which cannot be easily fitted into the logical arrangement and are normally the subject of clauses at the end of the General section of the Specification. They are likely to be specific to the type of work and the type of client involved in the contract. Three such miscellaneous clauses which are commonly required are:—

(1) The supply, erection, and subsequent removal of one or more site name boards, with a specified layout, giving the name of the Employer and the Contract, the names of the Contractor and sub-contractors and the names of consultants.

(2) The provision of progress photographs (usually monthly) taken by a professional photographer. The clause usually states the approximate number of photographs in a set and the number of each to be supplied, and gives requirements for the labelling of the photographs with description and position, for albums (if required), for control by the Engineer of the positions from which the photographs are taken, and for the ownership of the copyright in the photographs (including ownership of the negatives).

(3) Particularly in public authority contracts, the control of publicity by the Engineer or through him by the Employer's public relations department and the control of advertising at the site and particularly on Site hoardings.

5.8 Sub-Contracts

In most cases, a main contractor will sub-let part of the work using a standard form of sub-contract (either a national sub-contract form or a company form) which provides that the sub-contract work shall be carried out in accordance with the requirements of the main contract. The sub-contract specification then consists of the relevant parts of the main contract specification. Any special general or particular provisions are usually set out in a separate document or schedule referred to in the form of sub-contract. These provisions usually consist of a list of the facilities which will be made available by the main contractor and terms for their use, as well as the provisions to be made by the sub-contractor such as hutting, plant and materials store, welfare provisions, and other matters required under the main contract but not being provided by the main contractor.

In principle, the main contractor is entirely responsible for his relations with and the work of his sub-contractors and, therefore, the form and content of his sub-contract documents are a matter entirely for him. In practice, particularly with specialist sub-contractors employed on erection or installation at site, the main contractor often fails to satisfactorily supervise specialist sub-contractors (relying on the Engineer's supervision to ensure satisfactory work), except as it affects the programme for other items or trades, and often fails to fully inform the sub-contractor concerning the general contract responsibilities. This gives rise to problems during the course of the contract and results in friction between the Engineer's Representive, the Contractor and the sub-contractor. For these practical reasons the Specification often contains a clause defining, in general terms, the information which the main contractor must provide to his site sub-contractors. This clause may consist of a specific requirement in a particular section of the Specification (such as that set out in item (1) of Part 5.7.2 of this Chapter) or (particularly when there are a number of specialist sub-contractors) a general clause may be included in the General section. The following general clause has been used in a number of contracts and is typical of such a general provision –

> The Contractor shall ensure that all his sub-contractors are in possession of all those parts of the Specification and other Contract documents as are relevant to their work, not being limited to the specialist sections related to their expertise. In particular, but without prejudice to the generality of the foregoing, every sub-contractor shall receive copies of Specification Section... and be bound to carry out his work in accordance with the applicable clauses therein.

In Part 4.5.9 of Chapter 4 it is recommended that the section of the

Conditions dealing with nominated sub-contracts should include clauses setting out the procedure to be followed by the main contractor for obtaining quotations from suitable sub-contractors. Although these are most appropriately provided in the Conditions, they are not included in the national standard forms; furthermore the particular procedures usually vary according to the organisation responsible for preparing the contract document. For this reason the provisions for obtaining nominated sub-contract tenders during the course of the contract are often included in the General section of the Specification, rather than as an amendment or addition to the standard form of Conditions being used.

Similarly, the typical clause concerning pre-ordering which is given in Part 4.5.9 of Chapter 4 is often included in the Specification. Where only one type of material is being pre-ordered, then the clause dealing with that material and the reference in the typical clause to "these are described in the Contract and copies are annexed to the Contract" would be amended to "copies of these orders are annexed in the Schedule to this Clause."; where orders for a number of different types of materials or work are placed prior to letting the main contract, the typical clause would be in the General section of the Specification and references to "the Contract" should be amended to read "Schedules to clauses of this Specification". Where the typical clause is included in the Conditions of Contract or in the General section of the Specification then each of the technical specification clauses to which a pre-order schedule is attached must state –

> In accordance with Clause... of the Specification *[or Conditions of Contract]* a copy of the order for .... is attached as a Schedule to this clause.

CHAPTER 6

Payment Documents

6.1 Introduction

Chapter 4 (Parts 4.5.14 and 4.5.15) discusses the clauses to be included in the Conditions of Contract to enable the Contract Price and the Terms of Payment to be evaluated. Those clauses should set out the general items of work for which payment is to be made by the Employer to the Contractor, provide for any contra charges, and state the times at which payments and advances are to be made. The precise method of evaluating and calculating the amounts of the payments and advances is not usually given in the Conditions but is set out in accompanying Schedules or Bills of Quantities, which are often longer than the Conditions themselves and are not, therefore, suitable for inclusion in them, except by cross reference. The form and contents of the Schedules and Bills will vary with the type of contract and the character of the Works, but they may all be described as "Payment Documents" and will represent that part of the contract documentation which defines the financial provisions in detail.

As mentioned in Part 4.5.14, contracts can be classified (in respect of payment provisions) into two main types, described respectively as:

(a) Lump Sum contracts in which the Employer pays the Contractor a lump sum for the whole of the work initially described in the Contract; this lump sum is subject to adjustment in respect of variations in the Work and of contra-charges made in accordance with the terms of the Contract; or

(b) Measure and Value contracts where the Employer pays the

Contractor for the quantity of work carried out, as measured and valued in the manner defined in the Contract, subject to deduction of contra charges as provided in the Contract.

There may also be provisions in the contract for general claims, such as for disruption of work (resulting from various causes) for which methods of valuation may either be laid down in specific clauses of the contract or follow from a reading of the whole of the contract provisions.

Lump Sum contracts are the commonest form in building work and in mechanical and electrical engineering work and may be used in some civil engineering work where the quantities of the various types of work cannot be fixed or accurately defined initially (at tender stage, where there is formal tendering). For building works and for some civil engineering works the tendered lump sum is usually accompanied by detailed Bills of Quantities which define all the individual items of work, of material and equipment supply, and of general preliminary work which make up the Works. These are priced item by item and the total represents the tendered lump sum. The Bills of Quantities are used to value authorised variations of the Works to enable the tendered lump sum to be adjusted to arrive at the final contract sum, as well as being used for the valuation of work for interim payment during the course of the contract. Those Lump Sum contracts in which provision is made for interim payments and which do not include Bills of Quantities in the documentation (mainly contracts for mechanical and electrical engineering work) are likely to require a schedule defining in detail the manner in which interim payments are to be calculated or valued, which can be referred to as a "Payments Schedule". The documentation may also include a schedule setting out in detail the methods of valuing authorised variations, which can be referred to as a "Variations Schedule".

Most civil engineering contracts are of the Measure and Value type and include Bills of Quantities in the documentation. These Bills are used to value the work carried out, in order to arrive at the final Contract Sum; they are also used for valuations for interim payments. In this type of contract, authorised variations are automatically measured and valued in accordance with the tendered Bill items or with variations of Bill items fixed by the Engineer in the course of the contract. Many Measure and Value contracts, for mechanical and electrical engineering work (and also for civil engineering work) do not have Bills of Quantities but are valued in accordance with a Schedule of Rates which sets out unit rates of payment for various items of work or for the supply of material or equipment. From this schedule the value of the Contract Price is calculated by measurement of the quantities of the various items in the Works. Schedules of Rates should be accompanied by a schedule of approximate quantities of the main items of the work to enable the approximate total value of the contract to be calculated at the time of tendering.

The rates of payment in Bills of Quantities or in Schedules of Rates only apply to contract quantities or values of a similar order to those in the contract as originally entered into. If the contract quantities or values vary excessively, then the rates may have to be varied; the extent of variation acceptable before consideration is given to variation of the rate is often defined in the pricing document (or sometimes in the Conditions of Contract). It will usually be of the order of 15% to 25%, or occasionally up to 50%, of the total value of any item and/or of the contract.

6.2 Bills of Quantities

Bills of Quantities originated with building and civil engineering contractors who had to tender fixed prices in competition for various works. Each contractor would, from the drawings supplied by the client calling tenders, calculate and list the quantities of each of the various types of work, which he would then price from his experience of that type of work or prepare a price based upon a calculation of the amount of material and labour involved in a unit item. Each tendering contractor prepared his own Bill of Quantities (not disclosed to the client), which varied in content from one contractor to another; this system still prevails in the United States. In the UK, clients and contractors came to the conclusion that it was wasteful for Bills to be prepared many times over by the different tenderers for the same work, and the practice grew up whereby a Bill of Quantities was issued with the other tender documents, to enable tendering contractors to submit offers on a common basis without having to take off their own Quantities. This arrangement was possible because building and civil engineering work was carried out to designs prepared by the client's technical advisers and did not involve design by the contractor. The system did not evolve in electrical and mechanical engineering work because this normally included a substantial proportion of manufactured items involving design by the manufacturer contracting to do the work, so that the quantities of the various items making up the work varied according to the particular manu-facturer's design. Where the main part of the manufactured work was designed by the client's advisers (such as in structural steelwork, where the fabricator's design was usually limited to the joints) the system of Bills of Quantities was used. A disadvantage of the system of common Bills was the variety of ways in which measurements defining the work were made, so that tenderers were uncertain what was included in particular quantities. As a result, in the 1930's, the rules for preparing Bills of Quantities were standardised and codified for the different classes of work; documents called "Standard Methods of Measurement" (setting out these rules) have been in use for many years.

In the UK the documents currently in use which set out the rules for preparing Bills of Quantities are:—

The Standard Method of Measurement of Building Works, published by the Royal Institution of Chartered Surveyors.
The Civil Engineering Standard Method of Measurement, published by the Institution of Civil Engineers.
The Method of Measurement for Highway Works, published by Her Majesty's Stationery Office.

These standard methods of measurement are in each case related to a standard form of Conditions of Contract, i.e. the Building Method relates to the JCT Conditions, while the Civil Engineering Method and the Highway Works Method are related to the ICE Conditions. The measurement documents are not easily used with substantially different conditions of contract, e.g. the Civil Engineering Method is not easily applied in a contract which incorporates the JCT Conditions. For this reason, the Standard Methods are not applicable to international and to overseas contracts.To attempt to overcome problems arising in this connection, the Royal Institution of Chartered Surveyors has published a document entitled Principles of Measurement (International) for Works of Construction, but this has been little used in engineering work.

The contents and methods of preparation of Bills of Quantities have developed over many years and have become standardised, to the extent that "libraries" of item descriptions are available for drafting the Bill, and a substantial part of the operation of Bill drafting can be computerised. There is also an extensive literature on the preparation and drafting of Bills of Quantities to which reference can be made by engineers engaged in drafting Bills. Discussion of the drafting of the Bills of Quantities in this Chapter is, therefore, limited to the general principles which need to be understood in connection with the drafting of other payment documents, particularly by engineers not familiar with the preparation of Bills of Quantities.

Bills of Quantities consist of a schedule, defined as "a list of items giving brief identifying descriptions and estimated quantities of work comprising the execution of the works to be performed", and a Preamble which sets out the basis upon which the Bill has been prepared and is to be priced. The Preamble includes cross references to other documents such as the Conditions of Contract, the Specification and the Method of Measurement.

The Bill schedule is in a standard format consisting of columns with the following headings :—

Item	Description	Unit	Quantity	Rate	Amount

The first column is for the numbering of each item while the second column contains descriptions of items of work together with headings and sub-headings relating the items to relevant parts of the Works. Against each item is set (in columns 3 and 4) the unit of measurement and the quantity of that work, so that tenderers can insert the rate and the extended amount in columns 5 and 6.

The Method of Measurement gives the unit in which the measurement of each item of the Bill is to be made; this will be linear units or area units or cubic units or weight or time (hour, day or week) or number or, where none of these are applicable, a sum which is to be priced by the tendering contractor. In some cases (where the work cannot be accurately defined) a Provisional Sum is shown in the unpriced Bill; this is an amount to be included in all tenders.

The item descriptions are prepared in accordance with the rules in the Method of Measurement and are often taken from a standard library of descriptions. The quantities are calculated from the drawings and are usually measured net, any waste allowance being included within the rate for the work, in accordance with Method of Measurement rules. Provision is made for totalling each page and carrying it forward to sub-summaries for each Bill (referred to as "Collections") and a Grand Summary with total for all the Bills. The Methods of Measurement require that a separate Bill shall be written for each class of work. This format can be used for many other payment documents with the omission of inappropriate columns, e.g. where there are no quantities, the columns for "Quantities" and "Amount" would be omitted.

As mentioned previously, the standard methods often require amendment for particular contracts and these amendments are made in the Preamble. In some cases the standard is not appropriate and a method of measurement specific to the contract has to be appended to the Preamble. For those who are not familiar with Bills of Quantities, the contents of the Preamble are best explained by an example. The following is the Preamble for Bills of Quantities forming part of the documentation of a contract for a small tunnel. This tunnel was constructed as part of highway works; the Civil Engineering Standard Method of Measurement was not applicable, but the Method of Measurement for Highway Work did not include provision for tunnelling. A Method of Measurement for tunnels in highway works was, therefore, drafted and appended to the Preamble, and this is referred to. The complete preamble (which is based on the model Preamble in The Standard Method of Measurement for Highway Works) is as follows –

PREAMBLE

1. In this Bill of Quantities the sub-headings and item descriptions identify the work covered by the respective items, but the exact nature and extent of the work to be performed is to be ascertained by reference to the Drawings, Specification and Conditions of Contract as the case may be read in conjunction with the matters listed against the relevant marginal headings 'Item coverage' in the "Method of Measurement". The rates and prices entered in the Bill of Quantities shall be deemed to be the full inclusive value of the work covered by the several items including the following, unless expressly stated otherwise:—

    (i) Labour and costs in connection therewith.
    (ii) The supply of materials, goods, storage and costs in connection therewith including waste and delivery to Site.
    (iii) Plant and costs in connection therewith.
    (iv) Fixing, erecting and installing or placing of materials and goods in position.
    (v) Temporary Works.
    (vi) The effect on the phasing of the Works of alterations or additions to existing services and mains to the extent that such work is set forth or reasonably implied in the documents on which the tender is based.
    (vii) General obligations, liabilities and risks involved in the

execution of the Works, set forth or reasonably implied in the documents on which the tender is based.

(viii) Establishment charges, overheads and profit.

2. The measurement of work shall be computed net from the Drawings unless stated otherwise in the Method of Measurement. The Contractor shall allow in the rates and prices for waste.

3. Items against which no price or rate is entered shall be deemed to be covered by the other rates and prices in the Bill of Quantities.

4. Where in the Contract a choice of alternative materials or designs is indicated for a given purpose, the description billed and the rates and prices inserted shall be deemed to cover any of the permitted alternative materials or designs which the Contractor may elect to use and all measurement of such work shall be based upon the design to which those billed descriptions refer.

5. The information in the contract as to the whereabouts of existing services and mains is believed to be correct but the Contractor shall not be relieved of his obligations under Clause 11 of the Conditions of Contract. The Contractor shall include in his rates and prices for taking measures for the support and full protection of pipes, cables and other apparatus during the progress of the Works and for keeping the Engineer informed of all arrangements he makes with the owners of privately owned services, Statutory Undertakers and Public Authorities as appropriate and for ensuring that no existing mains and services are interrupted without the written consent of the appropriate authority.

6. Labours in connection with nominated Sub-contractors shall include:

(a) in the case of work or services executed, for affording the use of existing working space, access, temporary roads, erected scaffolding, working shelters, staging, ladders, hoists, storage, latrines, messing, welfare and other facilities existing on Site and the provision of protection, water, electricity for lighting, and clearing away rubbish and debris arising from the work;

(b) in the case of goods, materials or services supplied, for taking delivery, unloading, storing, protecting and returning crates, cartons and packing materials.

7. For the purpose of this Contract and unless otherwise more specifically defined in the Method of Measurement any curved

> work to a radius greater than 30 m is measured as straight. The Contractor shall allow in the rates and prices for taking measures required for constructing smooth curves to the radii shown on the Drawings.
> 8. For the purpose of this Contract the Method of Measurement referred to in Preamble 1 – General Directions is in accordance with the pages immediately following.

This preamble gives the cross references to other documents and detailed statements setting out the costs and risks which are to be included in the rates for all the items. The additional details of costs and risks which are specific to particular items are referred to in the Method of Measurement which, in this case, is appended to the Preamble. **The principles and the drafting of this preamble can be applied to many other pricing documents.**

6.3 Schedules of Rates

Although a Schedule of Rates appears to be a Bill of Quantities with the Quantities column and the Amount column omitted, the important difference between the two is that the Bill is intended to describe the whole of the Works so that the sum total of the Amount column represents the tendered price. A Schedule of Rates, on the other hand, is merely a basis for arriving at the Contract Price and may not include many small items, the price of which may be the subject of negotiation during the course of the contract. It is, of course, very desirable to include all the items large and small which make up the Works, to enable firm prices to be fixed at commencement of the contract, but the Schedule of Rates does give greater flexibility where there are limitations on knowledge of the extent of the Works at the initial stages. Tendering contractors need approximate quantities of the main items to enable them to rationally analyse the cost; at the very least they must have a knowledge of the approximate total costs of the Works or, in the case of annual maintenance contracts, the expected rate of working.

When drafting Schedules of Rates it is very important to ensure that the item description accurately defines the work included in the item concerned. Bills of Quantities usually rely upon a standard method of measurement and upon the definitions in the Preamble to ensure that the work included in the various items is accurately defined. In Schedules of Rates for civil engineering work this is often achieved by incorporation of one of the standard methods of measurement together with a preamble similar to that used for Bills of Quantities (on the lines set out above). This procedure would not usually be adopted in other branches of engineering where there are no standard methods of measurement nor is it essential in civil engineering, particularly where it is considered desirable to use item descriptions which include a number of classes of work in each description.

The definitions in a Method of Measurement allow the descriptions in the corresponding Bill of Quantities to be shortened. Where, as in the case of a Schedule of Rates which is not based upon any standard method of measurement, these definitions do not exist the draftsman must either write a method of measurement to be appended to the Preamble or, alternatively, extend the descriptions to incorporate the definition which would otherwise be given in the Method of Measurement. The latter method is often more appropriate for Schedules written for contracts where the standard Methods of Measurement are considered inappropriate, such as in the process industries and in other mechanical and electrical engineering work. In those contracts, where the Engineer often acts as a managing contractor, the payment document will frequently consist of a relatively short Schedule listing items of work which each include a number of trades (or classes of work) and a variety of materials and in which the risks and extent of work is less well defined or more diffuse, i.e. the description and the consequent rates and quantities are more approximate and, if they are to be considered fixed in price, require the parties to accept the kind of risks attached to such approximation. The Preamble must reflect this state of affairs; the following Preamble to a Schedule of Rates, which formed the payment document in one of the contracts for a process plant, illustrates the character of the drafting required –

PREAMBLE

This Schedule gives rates for the various classes of engineering work to enable the contract price as defined in the Conditions of Contract, to be ascertained from the measurements of the quantities of work to which the various rates refer.

Measurements of concrete work shall be nett the dimensions shown on the drawings. The space occupied by reinforcing steel will not be deducted from the volume of concrete, but holes and cavities will be deducted.

Reinforcement shall be measured nett off the drawings and for the purpose of measurement shall be assumed to have a weight of 0.78 kg/sq.cm/m.length.

Measurements of excavation, where not included in the rate for other works, shall be measured the nett area of the bottom of the excavation, as shown on the drawings or as otherwise instructed, multiplied by the average depth from the ground level at the time of commencing excavation to the bottom of the excacation, no allowance being made for slopes or strutting, the cost of which will be deemed to be included in the rate.

The rates for excavation shall be deemed to include the cost of excavating in dry or wet soil and of removing all water which may enter the excavations.

Measurements of filling, where not included in the rate for other work or otherwise stated shall be the nett volume to be filled, as measured from the drawings or, where this is not possible, measured at site, and no allowance shall be made for bulking, settlement of fill, consolidation, or similar factors.

Measurements of drains shall be the nett length from centre line of gulley or other connecting point to centre line of main drain or other connecting point, and no deduction will be made for the width of the gulley or the main drain.

Measurements of carbon steel pipelines to be the length of pipe on its centre line from inlet connection to outlet connection. The length of branch lines to be measured from line of main pipe to outlet; where the connection is between two pipes, from centre line to centre line of main pipes. Deductions of length are to be made where valves are inserted. The deducted length shall be the distance between faces of valve flanges or of valve ends on butterfly valves.

Unless otherwise specifically stated to the contrary, the rates in this schedule are deemed to include the cost of all constructional plant, materials and labour required for the finished work described under the respective items and all preliminary costs for supervision, offices, labour welfare, setting out, temporary works, clearing up on completion, and all other general liabilities, obligations and risks arising out of the Conditions of Contract and the Supplementary Conditions, and of the Specification.

The descriptions in this Schedule of Rates have been prepared upon the basis that the tendering contractor is experienced in the requirements of engineering in connection with petrochemical works. So far as possible, these descriptions indicate the various operations, the cost of which are included in the rate but, notwithstanding the omission of the description of any operation, the rates shall be deemed to be inclusive of all the costs of all operations which may be ascribed to the class of work referred to in the item, and no claim will be made in respect of alleged defective description on the grounds that further operations other than those described are necessary for the provision of the class of work required by an item of this Schedule.

The drawings attached to this Schedule for tendering purposes are preliminary only, and it may be anticipated that there will be alterations to the form of the work in the course of developing and finalising the engineering designs. The purpose of the preliminary drawings is only to indicate the general extent and type of the work, to enable a contractor experienced in petrochemical engineering work to ascertain the class and type of work to which the descriptions in the Schedule refer, and to ensure that the rates inserted include the cost of all the operations which may be involved in carrying out the work. Any explanations required by tendering contractors in connection with the class of work indicated on the drawings or described in this Schedule, will be provided prior to tendering and submission of an offer by the tendering contractor in the form of a complete Schedule of Rates with all rates inserted will be deemed to indicate that the Contractor has fully satisfied himself as to the requirements of the work.

In the course of development of the detailed drawings, the proportions of the various operations included within the items described in this Schedule may vary from the proportions which may be inferred from the drawings attached to the tender documents or from the Contractor's assessment of those proportions. The rates inserted in this Schedule are deemed to include provision for all risks and advantages which may accrue from this procedure.

Certain items included in this work, such as grouting, roads, trenching, etc., may require the Contractor to make available labour for such items, after the main work has been completed. It is intended that the Contractor shall make appropriate provision for reduction in the overall cost of supervision when the main works are completed, so that the items referred to bear an appropriate supervision charge. No claims in respect of increased cost of preliminary items such as supervision, office accommodation, welfare facilities, etc., will be made in connection with such items of work carried out after the main works are complete or of waiting upon other trades on a daywork basis, up to such time as the erection of the plant has been completed, and the plant handed over complete to the Employer.

It will be seen that the Preamble includes rules defining the extent of measurement of different types of work, these rules being applicable to a number of items. It also includes the necessary clauses covering the general definition of labour, materials and equipment to be included in the rates (similar to the preamble to Bills of Quantities) and the provisions required to bind the parties to a more approximate type of description and information. The Preamble does not include definition of the classes of work included in each item; these have to be set out in the item descriptions which are, therefore, considerably longer than would be expected in Bills of Quantities. An example of description (for underground piping), in the format of the Schedule, taken from the same document as the Preamble, is as follows –

No	Description	Unit	Rate
20	**Carbon Steel Piping** Take from store and install complete carbon steel underground piping, externally coated and wrapped with bituminous compound. And including all costs connected with carefully handling the pipes from store to position of installation so as to avoid any damage to the protective coating. And including the cost of fabricating all reducing pieces. Reducing pieces measured as straight pipe of size of larger end. And including the cost of excavating trenches for cooling water piping as specified, to a depth giving cover over pipe of 2'-6" and to such increased depths as may be necessary at pipe crossings. And including the cost of lowering and positioning the pipes in the trench and providing temporary supports.		

	And including the cost of supplying the Engineer with a welding procedure for approval and modifying such procedure, if necessary in accordance with the Engineer's requirements, and of qualifying welders to the satisfaction of the Engineer. And including the cost of welding piping in accordance with the approved procedure. And including all costs of making good the coatings at welded joints. And including all costs of testing the whole of the completed pipeline and of testing parts of the pipeline separately if the Engineer shall consider this necessary for constructional reasons and including all costs of making good faulty welds which may be disclosed during the test. Testing to be a pressure of 12 bar for cooling water piping or 15 bar for fire main piping. And including the cost of back filling with sand and soil and of consolidation as specified. Finishing with 300 mm of slag at ground level. (a) 50 mm pipe Schedule 40 (b) 100 mm pipe Schedule 40	m m	

The actual schedule includes pipes up to 1.2 m in diameter and this item is followed by items described as "Extra over" for the additional costs of fittings such as bends, tees, stub ends, etc.; a typical description for such Extra over items was as follows–

21	Extra over costs in item 20 for fabricating 90° bends including all costs of cutting pipe and welding and of preparing ends of pipe for welding into pipeline. (a) 50 mm pipe Schedule 40 (b) 100 mm pipe Schedule 40	Nr Nr	

The term Extra over refers to additional work or supplies in connection with an item previously measured and priced; the use of the term avoids deduction in measurement to allow for the additional work to be described separately, e.g. in this case the pipeline is measured its total length including length of bends and other fittings and the use of the term Extra over provides for the additional cost of the fittings, whatever their length. It is particularly convenient to use this term when the item concerned may vary in length, depending upon the particular manufacture or standard which the contractor adopts in his pricing. The abbreviation Nr is the standard abbreviation for a measured number which is adopted in the European Community and is not to be confused with the abbreviation No., used generally in English.

6.4 Payment Schedules

As discussed in Chapter 4 Part 4.5.15, most engineering contracts provide for interim payments monthly during the course of the contract or stage payments when specific stages of the work have been completed. These interim payments are usually made against the collateral of completed work which has become the property of the Employer; the contract must, therefore, provide for the valuation of the completed work month by month. Where the contract documents include a Bill of Quantities or a Schedule of Rates, evaluation for interim payments is relatively straightforward and consists of approximate measurement of the work, priced at the rates in the Bill or the Schedule. Any work not described in the Bill or the Schedule is then priced at rates fixed by the Engineer (often the rates are provisionally fixed for interim payments and finally fixed on completion) and the resulting valuation certified by the Engineer.

In the case of lump sum contracts for mechanical or electrical work or for "design and build" contracts there may be no Bill of Quantities or Schedule of Rates: valuation for interim payments will usually be made based upon a schedule of the proportion of costs of various stages or parts of the work, either in the form of percentages totalling 100% or in the form of lump sums which total to the tender price. Where the contract provides for monthly interim payments, the Engineer will base his monthly valuation upon the approximate percentage of each part or stage of the work which has been completed at the end of the month and will certify accordingly. Where only stage payments are to be made then the Contractor will only be entitled to be paid against the Engineer's certificate that the stage concerned has been substantially completed (full completion down to the last nut and bolt may not be practicable because of connection to other parts or for other similar considerations).

When design is to be carried out by the contractor who is to construct the Works, then either such a Payment Schedule setting out the proportion of the contract price for each stage should be included in the tender document or, alternatively, the successful tenderer should be required to provide a Bill of Quantities or other definitive payment document before the contract is awarded. The latter alternative is only feasible if adequate time is allowed for the tenderer to prepare a reasonably detailed design. Any payment document prepared by a tenderer must be carefully examined and any necessary revision negotiated, to ensure that it truly represents the work proposed by the tenderer, so as to avoid the risk of substantial front end loading which would result in early payment being excessively high in relation to the amount of work actually done at the time. In extreme cases it might be necessary to redraft the document in agreement with the tenderer. An alternative is for the Contractor to provide on-demand bonds against interim payments and thus avoid valuation of the work each month, each interim payments being a fixed percentage of the contract price and the on-demand bond being increased prior to the Employer making payment, so that the total of the interim payments is always covered by the bond. This method might add about 1% to the cost of the Works to pay bank charges for the bonds.

If payments are to be made in very few stages (such as is common in shipbuilding), the stages and their percentages may be included in the Conditions of Contract. Otherwise, a Payment Schedule should be included in the contract. This would be in the form of a simple preamble setting out the method of valuation, followed by a schedule setting out either the price of each item or, preferably, the percentages of the various items of work; the schedule should total either one hundred per cent or be equal to the tender price. As an example, a schedule which was used for a building within a process plant, is given below. In the example the actual percentages have been omitted.

INTERIM PAYMENTS SCHEDULE
PREAMBLE
Interim payments will be made according to the value of the various parts of the work completed in accordance with the percentages set out below. Partially completed sections of work shall be paid for in accordance with the proportion of the appropriate percentage estimated approximately by agreement between the Contractor and the Engineer.

Tendering contractors shall insert in this Schedule the cost percentages of the various items as a percentage of the tendered Lump Sum. This percentage shall be based upon all preliminary and general costs being spread over all items as a common percentage addition to the nett cost of labour, materials and plant. With the approval of the Engineer, additional payment may be claimed in respect of materials at Site in accordance with an estimate agreed with the Engineer.

Schedule

Item	Description	%
1	Cast iron drainage goods	..
2	Manholes	..
3	Concrete work in Ground Floor	..
4	Concrete work in walls and pilasters of building, including stairs	..
5	Concrete work in First Floor of building, including slab over control room and stairs	..
6	Concrete work in foundations of tank	..
7	Concrete work in foundation to process tower and in foundation to emergency stairs	..
8	Concrete work and brick walls for pits	..
9	Brickwork in building	..
10	Plastering in building	..
11	Screeding in building	..
12	Doors and windows and all other building work	..
		100

| 13 | P.C. sums amounting to £25,000 plus ..%. Payment to be in accordance with payments actually made by the Contractor to supplier or subcontractor, plus this percentage. | £........ |

Although this schedule was prepared for building work the principles apply equally to other types of construction work. Where the contract consists of the manufacture of equipment and its erection at site, then the last paragraph of the Preamble will usually provide for payment for equipment only when delivered to site or to the Employer's store, but not intermediately during manufacture, unless satisfactory arrangements are made for legal ownership of the manufactured material to pass to the Employer. A suitable alternative paragraph might read as follows –

Payment for items of equipment will only be made when such equipment has been delivered to Site or into a store provided by or approved by the Employer and such equipment is clearly marked as the property of the Employer.

The schedule given above is reasonably simple but sufficient for its purpose. Because the percentages and the payment are necessarily very approximate, it is essential that the draftsman does not introduce such complexity as would give the appearance of greater accuracy than is justified, in order to avoid the possibility of it being contended that the percentages set out in the Schedule accurately represent the cost of the various stages or items. Such a contention might, if sustained, lead to the percentages being used as the basis for negotiations concerning claims and variations; the nature and the manner of calculation of the percentages is not suitable for such a use. If, instead of percentages, sums are to be inserted, then this will increase the apparent accuracy of the schedule by an order of magnitude and it may be desirable to extend the preamble in order to define the items more fully; as an example, sums in a small contract of £100,000 would be given to the nearest £10, but percentages would be given to the nearest 0.1, representing £100.

6.5 Payment for Variations

Even in relatively straightforward contracts incorporating Bills of Quantities, payment for the work involved in carrying out the Engineer's variation orders may give rise to serious disputes. It is, therefore, important to include, within the contract documentation, as much data concerning the prices for various items of the work as will enable as many as possible of the variations to be valued at rates set out in the contract.

As mentioned in Part 6.1 of this Chapter, where Bills of Quantities or Schedules of Rates form part of the contract documentation, the rates and prices in these will normally be used for valuing those variations which consist of work similar to the items in the Bills or the Schedule. Dissimilar items and the effects of any disruption to the contract work will still have to be negotiated. In the case of Lump Sum contracts where there are no Bills or Schedule of Rates it is highly desirable that a document be included which will provide a basis for valuing foreseeable types of variations. This most commonly occurs where the Contractor is responsible for design as well as construction and where it is not feasible to provide detailed Bills or a Schedule of Rates at the stage of tender acceptance because a sufficiently detailed design is not then available.

In such cases the form and drafting of a Variations Schedule will have to be closely related to the contractual provisions and the types of work in each case. In the case of construction contracts based upon modifications of one of the accepted standard forms of Conditions, the Variations Schedule can often consist of a schedule of lump sums for each of the different classes of work to be carried out, accompanied by a preamble which provides that variations shall be valued proportional to the measured amount of the principal item of that class of work which would form part of the final detailed design of the unvaried item. Thus, where the work includes different classes of pipework, the length of pipe of each class of pipework can be determined for the unvaried detail design. Variations involving changes in the length of different classes of pipework can then be valued in proportion to the lump sum for that class of pipework, provided that the document preamble states that secondary items such as joints and other fittings, brackets, insulation, corrosion protection, and the like, are included in the measurement of the principal item, i.e. the main pipe of that class of pipework.

Such a variation schedule can be conveniently combined with an Interim Payments Schedule which gives lump sums for each of the items. As pointed out in the previous Part of this chapter, the items will need to be more fully defined than would be the case for a schedule which merely included percentages for each of the items, rather than lump sums.

Such schedules can only deal with variations in the extent of the types of work listed and then only within specified limits. The limitations which apply to the schedule, such as the extent of variation in total quantity which is acceptable for pro-rata adjustment and the provisions in respect of equipment supply and erection, must be included in the preamble. A useful example of this type of schedule is given in the next Part.

6.6 Example of Variation and Interim Payment Schedule

The following is an example of the drafting of a combined interim payment and variation schedule which should assist the reader in his understanding of the previous two Parts of this chapter and may prove useful to draftsmen preparing this type of document for a design and construct contract.

SCHEDULE FOR INTERIM PAYMENTS AND VARIATIONS

PREAMBLES

1. The price set against each of the items in the Schedule represents the inclusive price for all the work in that item; the total of all those item prices is the tendered Lump Sum.

2. The item descriptions identify only the principal items of work involved. The item price includes all related secondary items and operations (such as, but not limited to, framing, fixings, brackets, finishes and finishings, joints and connections and items required for them, controls and instrumentation not otherwise itemised, and the like) so as to ensure that the prices comprehensively include all secondary items and operations referred to or reasonably to be inferred from the Conditions of Contract, the Specification and the Drawings in respect of the related principal items (whether or not they have been described in the Schedule as included within that principal item) so that the tendered Lump Sum derived therefrom and inserted in the tender represents the full inclusive value of the obligations of the Contractor for design, construction, completion and maintenance of the Works. In certain cases some secondary items and operations are included in the descriptions of the principal items to the extent necessary to enable such secondary items and operations to be correctly allocated, but the inclusion of such secondary items in a description shall not limit the comprehensiveness of the corresponding price nor exclude any secondary items referred to or reasonably to be inferred from the aforementioned documents.

3. Each of the items shall include:—

(i) Design and costs in connection therewith.

(ii) Labour and costs in connection therewith.

(iii) The supply of materials, goods, equipment, storage and costs in connection therewith, including but not limited to waste and delivery to Site and to and from all other places where such delivery is required.

(iv) Construction plant and costs in connection therewith.

(v) Demolishing or otherwise extracting and removing from the site existing structures and equipment and other items or materials.

(vi) Fixing, erecting, installing or placing materials, goods and equipment in position, including connected operations such as, but not limited to, fabrication, testing, commissioning, and the like.

(vii) Temporary Works.

(viii) The effect on the phasing of the work of alterations or additions to existing services and mains, and of limitations of occupation of various parts of the Site and of the various programme limitations set out in the Contract.

(ix) General obligations (including, but not limited to, facilities for the Engineer, as built drawings, operation and maintenance manuals and progress photographs), liabilities and risks involved in the design and execution of the Works set forth or reasonably implied in the documents on which the Tender is based.

(x) Establishment charges, overheads and profit.

4. Each valuation for interim payment shall comprise, as at the valuation date:

(a) the prices of all items in the Schedule which have been completed at the Site;

(b) a proportion of the prices of all items in the Schedule which have been substantially completed at the Site;

(c) a proportion of such items in the Schedule which have been partially completed at Site and which partially completed items the Engineer considers would be viable if the contract work ceased at the valuation date;

(d) 90% of the value of materials or equipment which is required for incorporation in the Works and which has been brought to the Site or has been stored in a manner and in a store approved in writing by the Employer.

The proportions referred to in the foregoing sub-paragraphs (b) and (c) shall be determined by the Engineer whose decision for the purpose of interim valuation shall be final and binding on the parties.

Rules for Valuation of Ordered Variations

1. In accordance with the provisions of the Conditions of Contract, the Schedule shall be used for valuing ordered variations where such method of valuation is appropriate in accordance with the following rules.

2. The pro-rata valuation of variations of items other than those having units specified as "sum" shall be determined from the formula –

$$V = P \times q \,/\, Q$$

where

V is the value of the variation,

P is the price of the relevant item as stated in the Schedule,

Q is that quantity (measured in the units specified in the Schedule) of the principal item whose price is P, which would have been incorporated in the Works in accordance with the Contractor's final design, if no ordered variation had been made,

q is the quantity of the variation measured in the same manner and the same unit as Q.

3. Where an ordered variation requires that an item which forms part of the Contractor's design and which has units specified in the Schedule as "sum" shall be varied, the value of the variation shall be limited to the net increase or reduction of cost of the material and/or equipment incorporated in the item including delivery to the Site. The net cost shall be the price paid by the Contractor for the materials or equipment delivered to the Site after deduction of all discounts, but including the cost of unloading and subsequent handling. Where, in accordance with an ordered variation, materials or equipment differing from those in the Contractor's "final design" are required then these rules shall not apply and the item shall be otherwise valued in accordance with the provisions of the Conditions of Contract.

4. Unless otherwise agreed by both the Contractor and the Engineer, these rules shall apply to variations of any item in the Schedule which do not result in a variation of the price of that item by more than 20%.

5. The Contractor and the Employer accept the risks and advantages of the provisions in these rules and recognise that the price adjustments which result will be approximate but acceptable.

These preambles and rules would be followed by a schedule of all the main items of the Work. A full schedule for any substantial work will be fairly lengthy; an example of part of a schedule, covering a number of typical items in certain types of construction contracts might be as follows.

Item	Unit	Description	Price £
		Foundations	
		Note: Excavation items shall include weak concrete blinding under foundations. Concrete items in foundations shall include pockets for and fixings to holding down bolts as well as all integral and surface finishes on other than formed surfaces. Formwork items shall include finishes to formed surfaces. Reinforcement items shall include all cover blocks or spacers and all chairs and other reinforcement supports.	
1	m³	Excavation for foundations.	
2	m³	Concrete in foundations.	
3	m²	Formwork to concrete foundations.	
4	t	Reinforcement in foundations.	
		Steelwork	
		Note: Steelwork items shall include all connections, stiffeners, gussets, shear connectors, bolts, studs, and other fasteners, holding-down bolts and base plates, brackets and the like and shall include all corrosion protection coatings, painting and other surface treatment. Light steelwork shall comprise stairs, ladders, handrails, open grid and chequered plate flooring and shall include all steel seatings, curbs, upstands and kicking plates.	
5	t	Steelwork in structural frameworks	
6	t	Light steelwork	
		Cabling	
		Note: Cabling items shall include conduit, trunking, cable trays, supports, earthing, joints, glands, all fittings for the foregoing, and connecting to switchgear and equipment.	
7	m	Power cables	
8	m	Lighting cables	
9	m	Control, alarm and instrument cables	
10	m	Communications cables	

		Pipework and Ductwork
		Note: Pipework items shall include joints and other fittings, supports, insulation, and connections to equipment (including valves) and supply mains.
11	m	Hot and cold water pipework
12	m	Fire protection pipework including sprinkler heads
13	m	Instrument pipework
14	m	Compressed air pipework
15	t	Ductwork including bends, tees, joints, tapers, flexible connections, smoke doors (including fusible links) vanes and dampers (both fixed and movable), sound attenuation, grilles and registers, blanking off connections to ductwork provided by others, supports, connections to equipment, and including thermal and acoustic insulation.
		Equipment
16	Sum	Lighting fittings, including steel standards, brackets, and lamp control gear.
17	Sum	Heating, ventilating and air conditioning equipment
18	Sum	Water supply and water heating equipment
19	Sum	Fire protection equipment
20	Sum	Security equipment and alarms not included elsewhere

This schedule is based on items extracted from a working document, suitably modified, and is intended merely to illustrate a particular method of dealing with the valuation of some of the ordered variations which may arise when the Contractor is responsible for both design and construction. No definition is given of the method of measuring the various quantities, as this is not important so long as the variation is measured in the same way as the original item quantities, as provided in paragraph 2 of the Rules. In case of dispute it might be necessary for the Contractor to produce his original quantities as evidence. Paragraph 2 of the Preamble provides that the price of the principal items shall be comprehensive but does not define which secondary items are to be included in particular principal items. This is dealt with in the body of the Schedule by notes covering the various groups of items; where only one item is involved (as in the case of Ductwork) then it is more convenient to include the definition in the description of the item.

6.7 Dayworks

Many variations comprise additional items the value of which cannot be satisfactorily estimated due to uncertainty or disagreement concerning the extent and/or the nature of the work. This applies to both remeasured and lump sum contracts and is dealt with by cost reimbursement in accordance with a Dayworks Schedule.

The Dayworks Schedule provides a list of rates to be paid for the various classes of labour, materials and plant used in dayworks, either in the form of a comprehensive list in the payment document or by reference to a standard Schedule published by an industry body. Probably the most useful schedule for construction work in the UK is the "Schedules of Dayworks carried out incidental to Contract Work" published by the Federation of Civil Engineering Contractors (FCEC). This consists of 4 schedules, 3 of them (Schedules '1. Labour', '2. Materials', and '4. Supplementary Charges' for items such as transport and subcontracts) give definitions of net cost and state and define percentage additions to be made to those net costs, while the fourth (Schedule '3. Plant') comprises an extensive list of construction plant items (in 1986 it was 26 pages in length) with their respective hire rates together with definitions of the cost items included in those rates. Although the FCEC Schedule is designed primarily for use in civil engineering work, it is written in a way which makes it applicable to a wide range of other engineering work. As an example, the definition "Amount of wages" provides for wage payments to be "...in accordance with the Working Rule Agreement of the Civil Engineering Construction Conciliation Board of Great Britain Rule Nos I to XXIII inclusive *or that of other appropriate wage fixing authorities current at the date of executing the work or where there are no prescribed payments or recognised wage fixing authorities, the actual payments made to the workmen concerned.*"(author's italics), thus permitting it to apply to other than civil engineering contracts. It is usual, when calling tenders, to give tenderers the opportunity to vary the percentage additions given in the FCEC Schedules as well as to add any plant items which may be required but are not included in 'Schedule 3. Plant'. To ensure that the Tender Sum reflects the amended Dayworks Schedule percentages, Provisional Sums (to which the amended percentages are applied) are included in the payment document.

The format of the Dayworks Schedule included in the contract will depend upon the format of the payment document, i.e. whether it is a Bill of Quantities or a Schedule of Rates or a Schedule of Prices or is in some other format. When tenderers are required to price individual rates for labour, materials and plant, all the items are listed in the Payment Document under a heading of Dayworks. When they are required to price adjustments to standard Schedules (such as the FCEC Dayworks Schedules) which are to be used for evaluating Provisional Sums, then a description item referring to the standard Schedule must be drafted, together with accompanying provisional sum items and items for percentage variations which are to be priced by the Contractor. A model draft for the use of the FCEC Schedules would be as follows:—

DAYWORKS

Daywork ordered by the Engineer pursuant to Clause ... of the Conditions of Contract shall be paid for at the rates, prices and percentage additions and under the conditions contained in the "Schedules of Dayworks carried out incidental to Contract Work" (referred to as "the Schedules") issued by the Federation of Civil Engineering Contractors current at the date of execution of the daywork, subject to the additions or deductions set out in the following items.

Labour
1. Provide the Provisional Sum of £......... for Labour in dayworks in accordance with the Schedules.
2. Addition/deduction* of ..%† to/from* item 1.
3. Provide the Provisional Sum of £......... for Materials in accordance with the Schedules.
4. Addition/deduction* of ..%† to/from* item 3
5. Provide the Provisional Sum of £......... for Plant in accordance with the Schedules *and the additions to Schedule 3 set out herein*
6. Addition/deduction* of ..%† to/from* item 5
7. Provide the Provisional Sum of £......... for Supplementary Charges in accordance with the Schedules.
8. Addition/deduction* of ..%† to/from* item 7.
   *Appropriate deletion to be made by Tenderer.
   †Percentage to be inserted and extended by the Tenderer. Deductions to be shown in parentheses.

The following items shall be added at the end of Schedule 3 – Plant; rates for all these items shall be inserted by the Tenderer.
[When additional plant items are required, the draftsman should incor-

porate the wording shown in italics and add the required items in a similar format to that adopted in Schedule 3, with item numbers beyond those currently in Schedule 3]

One of the advantages of using the FCEC Dayworks Schedules is that it is updated when price movements or changes in legislation or labour practices take place. This avoids disputes concerning the financial effects of variations ordered late in the Contract if substantial price movements have taken place in, for example, items such as oil (when oil prices were fluctuating wildly, amendments were issued several times in the year). The disadvantage for contracts which do not involve a substantial amount of civil engineering is that the percentage addition to wages will be varied when changes take place in civil engineering, even though those changes may not affect other trades.

Where the FCEC Schedules of Dayworks is not available (as in international or overseas contracts), the Dayworks Schedule will consist of:—

(i) Items for each class of workmen, which must be priced by the tenderer.

(ii) A Provisional Sum for materials and a percentage to be inserted by the tenderer to cover all the additions set out in the Preamble to the Payment Document.

(iii) Items for each type and size of construction plant. Tenderers will have to price all these items.

An alternative to (i) is to have Provisional Sum for net labour cost and a percentage (to be inserted by the tenderer) to cover all oncosts. This alternative will involve the Engineer's staff in checking the Contractor's wages sheets.

6.8 Cost Reimbursible Contracts

When work has to be started before all the requirements are known or before costs can be satisfactorily estimated or where a contractor will have to make special efforts to meet completion dates or to overcome problems which cannot be easily assessed, then a contract may be entered into whereby the Employer reimburses the Contractor his actual costs of carrying out the Works plus a percentage addition or a lump sum fee. The Conditions of Contract for reimbursible contracts are generally similar to those required for other contracts, but the payment clauses have to set out the percentage or the lump sum addition to the cost of the Works as well as the accounting and auditing systems and the interim payment provisions. In addition, there will have to be a full definition of the items to be included in the cost of the Works. Although this could be included in the definition clause of the Conditions, it will usually be sufficiently lengthy to merit either an additional clause or an Appendix or it may be included in a payment document.

When drafting the cost definition, the draftsman should take account of the following matters, among others:—

Which terms of employment of the Contractor's labour and staff are to be included in the cost;

Are advertisements and other expenses for engagement of labour and staff to be included in costs;

Discounts obtainable when purchasing materials;

Responsibility for waste of materials;

Payments for plant and tools owned by the Contractor and used on the Works;

Costs of rectifying errors made by the Contractor's staff and/or workmen;

Are insurances and financing costs to be included in cost of the Works;

Provisions for the disposal of surplus materials and plant during the course of the contract and on completion.

Any expenditure by the Contractor which is not included in the Cost of the Works will have to be covered by the percentage or lump sum addition. The definition of the Cost of the Works therefore also defines, by implication, what is included in the percentage or lump sum addition. That addition is usually intended to pay for the Contractor's Head Office overheads and profit, but disputes can arise as to whether, for example, the cost of visits by H.O. personnel to the Site are part of the Cost of the Works or covered by the addition. A similar cause of dispute can arise in connection with hire charges for the use of construction plant owned by the Contractor; should these contain a profit element or are all profits to be covered by the addition?

Although the definition of the Cost of the Works will vary from contract to contract, the main items are likely to be fairly standard. The following example, based on a number of past contracts, may be of assistance to draftsmen.

COST OF THE WORKS

The actual cost of the Work shall be all costs properly incurred for the purpose of the Works, as defined below:—

1. The wages and salaries of all supervisory staff and labour and others employed in the Works, together with other payroll costs and expenses including inter alia transportation, removal, welfare facilities and other allowances and costs which conform to the Contractor's regularly established practices concerning his employee's terms of employment, as well as all payments required by law, and the net cost of advertisements for staff and labour.

2. Fees and expenses paid to other organisations in repect of design, testing, experimental and other engineering services directly relating to the Works.

3. The cost of all materials (including fuel), whether permanent or temporary materials, when they are incurred, subject to the deduction of all reductions, rebates or discounts which can be obtained by the Contractor.

4. The cost of any machinery, plant, transport, scaffolding or tools purchased by the Contractor for the Works, subject to the deduction of all reductions, rebates or discounts which can be obtained by the Contractor.

5. The cost of hire of any mechanical plant or of any non-mechanical plant or equipment owned by the Contractor or by any subsidiary or by any group or company in which the Contractor has a controlling interest at rates published by a plant hire and/or an equipment suppliers' association approved by the Engineer, or at analogous rates to such published rates where specific rates for a particular item are not included in such publications. All subject to a reduction of 10% before inclusion as part of the Cost of the Works.

6. The cost of hire from external sources of any mechanical or non-mechanical plant or equipment at the invoice value of this hire, subject to all reductions, rebates or discounts which can be obtained by the Contractor.

7. The cost of carriage of all mechanical and/or non-mechanical plant or equipment on to the Site and their removal when no longer required, and the cost of any offsite repair or maintenance except insofar as the cost of such repair or maintenance is covered by hire rates for such mechanical or non-mechanical plant or equipment.

8. All payments to subcontractors approved by the Engineer.
9. Temporary offices, stores and compounds erected for carrying out the Works on or adjacent to the Site or at other locations instructed or approved by the Engineer.
10. Fees and other charges paid to local or other authorities and statutory undertakers.
11. Premium and other contributions for the insurance of the Works and all other insurance liabilities required in accordance with the Conditions of Contract where not elsewhere covered in this definition of the Cost of the Works.
12. The costs of visits to the Site by the public, VIP's and the like, when those visits are in accordance with an instruction by the Engineer or approved by him prior to such visit.
13. Bonds or guarantees obtained on the instructions of or at the request of the Employer or the Engineer and such bonds or guarantees in respect of site personnel as are obtained in accordance with the regularly established practices of the Contractor.
14. All other costs properly incurred for the purpose of the Works which are approved by the Engineer either previously or subsequently to their being incurred, as well as minor payments for miscellaneous services.
15. In the event that the Contractor shall sell or otherwise dispose of any materials or plant or other items used for the purpose of the Works, any actual or notional income resulting from such sale or disposal shall be credited in determining the Cost of the Works.
16. The Cost of the Works shall not include any costs arising as a result of –
    (a) deliberate or irresponsible acts of negligence by the Contractor or his servants or agents, and/or
    (b) any contravention by the Contractor or his servants or agents of any law or statute or of any Statutory Instrument or Regulation made thereunder.

When the addition is a percentage, no further payment document is required; profit and overheads on any variation in the cost of the work will be covered by a comparable variation in the sum calculated for the percentage addition. If the addition is a lump sum, then variations in the Works which are made by the Employer (either directly or through the Engineer), and which result in an increase in the cost of the work, may give rise to claims by the Contractor for an increase of the lump sum.

Lump sum additions may be either a fixed sum irrespective of the cost of the work or a sum which reduces if the Cost of the Works exceeds a target figure and increases if the Cost is less than the target, the latter arrangement being known as a Target Price Contract. Obviously, if the Employer makes changes to the extent of the Works which result in an increase in the Cost then the Contractor is likely to claim an increase in the Target Price and a corresponding increase in the lump sum addition. It is, therefore, necessary to include in the contract a detailed estimate of the Target Price or (in the case of a fixed lump sum addition) of the expected cost of the Works. There should also be a provision whereby the lump sum may be varied up or down if the extent of the Works is increased or reduced by the Employer by more than a specified amount, but not if the variation is due to natural or other causes provided for or implied by the Contract. It must be appreciated that cost reimbursible contracts are not necessarily free of payment disputes. A prudent draftsman will, therefore, safeguard the contracting parties by including cost definitions and estimates as appropriate.

CHAPTER 7

Miscellaneous Documents

7.1 Other Contract Documents, Bonds, and Guarantees

Previous chapters have dealt with the drafting of Conditions of Contract, Specifications and Payment documents. These form the main bulk of the documentation in engineering contracts, but the contract also requires an offer or tender from the Contractor and an acceptance by the Employer, these being often confirmed by a formal contract Agreement. In addition, many contracts require the provision of a Performance Bond and there may be requirements for on-demand bonds against advance payments and for retention, as well as guarantees by corporate groups where the contract involves a subsidiary or a consortium of subsidiaries of group members.

In theory bonds may be entered into by an individual, but in practice they are invariably given by a bank or insurance company. Contractors have regular arrangements for the provision of bonds by these organisations, and Employers are usually confident that such banks and insurance companies will have adequate resources to meet the bond if it should be called. The author has never heard of any modern contract which has been guaranteed by an individual, although some standard forms of bond still make provision for this eventuality.

Although the bonding bank or insurance company enters into an agreement with the Employer to pay him certain sums if there is a failure by the Contractor to comply with particular provisions of the Contract, the bond agreement does not normally provide for any payment or other consideration to pass from the Employer to the bonding organisation, even though the Contractor makes a payment (by a separate agreement) to the bonding organisation and the Employer may reimburse that payment to the Contractor. In order to make the agreement between the bonding organisation and the Employer enforceable in law it will, in the absence of consideration, have to be in the form of a deed. Section 20 of the Solictors Act 1957 makes it a criminal offence for anyone who is not a solicitor to draw up a deed in the course of business (i.e."in expectation of any fee, gain or reward") and engineers cannot, therefore, prepare a deed for a client although they could prepare a draft for a solictor to draw up a deed. Although this only applies in English law, it is advisable for engineers to avoid drafting bonds or similar documents where they lack special expertise, particularly in international contracts where clients may have special local legal requirements.

To avoid the use of deeds, some oil companies arrange to include a nominal consideration (say £10) in all bonds, to be paid by them directly to the guarantor.

In the UK it is not uncommon for clients to require a performance bond to be provided by a successful tenderer, although large public authorities often consider pre-qualification of proposed tenderers more economical; they have found it expensive to spend substantial sums on reimbursing contractors for the cost of bonds which are very rarely called. When a form of performance bond is required to accompany tender documents, it is advisable for the draftsman to adhere to one of the standard forms, such as those which accompany the ICE or the FIDIC Conditions.

In recent years it has become the practice among large contractors to split their business into a main holding company with a group of subsidiary companies each operating in a particular branch of contracting (such as civil engineering, mechanical and electrical services, process plant construction, etc.) as well as subsidiary companies for different parts of the country. Each of the subsidiary companies would have a much smaller capitalisation than the parent company. For large works it is common for the subsidiaries of a number of different groups to combine into a joint venture or consortium, the organisation being effectively a partnership of limited companies rather than a separate company. In these circumstances some clients believe that a contractor who is a group subsidiary or a consortium of subsidiaries of different groups should have their performance guaranteed by the holding company or companies of the group or groups. The form of such a guarantee would be that of an agreement under hand (i.e. not a deed), it being based upon the consideration that the guarantee would be given by each of the holding companies in return for the award of a contract to its subsidiary company, provided that it is entered into before the contract is awarded. Generally such group guarantees are only required in connection with large works where the Employer is a public authority or a large private company. In such a case the form of guarantee will normally be prepared by the Employer's legal department. It is preferable for engineers to avoid the drafting of such documents in which they have no special expertise, but if the form has to be drafted, then it should include the following matters:—

(a) The guarantor company shall indemnify and keep indemnified the Employer against the consequences of any failure by its subsidiary company to fully perform the requirements and provisions of the Contract.

(b) The liability of the guarantor shall not be greater than its liability if it had itself been responsible for the performance of the Contract.

(c) If the subsidiary company fails to perform and the Employer terminates the Contract and requires to enforce the guarantee, then the guarantor is to be given the opportunity of completing the Works in the terms of the Contract.

(d) No alteration in the subsidiary company's obligations nor any arrangement with the subsidiary company nor any forbearance of any of the actions of the subsidiary company should release the guarantor from his obligation.

If the client wishes to have the guarantee in the form of a deed, so that it will be effective for a period of twelve years (instead of six years for a contract under hand) then it must, of course, be drawn up by a solicitor.

7.2 Tender Forms

Most engineering contracts are based upon or refer to a formal tender by the Contractor; even where the terms of the contract have been negotiated between the Employer and a single contractor it is common for the contractor to make a formal offer on conclusion of the negotiations so that the contract is based upon an offer and an acceptance, with stated terms of payment, to enable the contract to be either under hand or in the form of a deed. A large proportion of engineering contracts are the results of competitive tendering based upon documents issued by the Employer which include a standard Tender Form for submission of the formal tender. A number of standard Conditions have tender forms annexed to them (ICE Conditions, FIDIC Civil and FIDIC M & E Conditions) which relate specifically to those Conditions and in each case there is an Appendix (which forms part of the Tender) setting out specific particulars of matters which are dealt with only in a general way in the Conditions. An alternative method of providing the information is adopted in the GC/Works/1, which requires a document described as 'Abstract of Particulars' to be included in the contract documentation. This document would normally contain the amendments and additions required to the General Conditions as well as any other special requirements and the specific particulars which are required in connection with matters dealt with only generally in the Conditions of Contract. No appendix to the Tender Form is, therefore, required where an Abstract of Particulars is provided. The information to be given in the appendix to the Tender Form or in the Abstract of Particulars will, of course, depend upon the requirements of the particular contract and of the Conditions of Contract used.

There are occasions when it is necessary to amend the tender documents during the tender period, either because further information has been made available (such as revised data on ground conditions or changes to the plant which is to be installed) or to developments in the client's requirements. There may also be requests for clarification from tenderers or errors may have been discovered in some of the documents. In the case of competitive tendering it is essential (in order to avoid complaints or objections from tenderers) that any amendments to the tender documents should be notified in writing concurrently to all tenderers, this notification being referred to as an amendment letter or a circular letter. The amendment letter should be drafted as an instruction to the tenderer to amend each of the documents referred to and should set out the specific wording of the amendments to particular sections, paragraphs and sentences or lines in each of those documents. In the case of drawings, the previous drawing number and revisions suffix should be "cancelled and superseded", the new drawing number and revision suffix "substituted" and copies of the revised drawing attached. The amendment letter should be accompanied by a receipt form and an instruction to the tenderer to return the receipt as soon as possible, as acknowledgement that the amendment letter has been received and will be incorporated in the tender. At the time when the last amendment letter is issued or at the end of the period available for issuing amendment letters, a revised Tender Form should be issued (under cover of an amendment letter) incorporating a statement that the tender documents have been amended in accordance with amendment letters listed and that the tender is based upon the amended documents.

The wording of the Tender Form will vary considerably, depending upon the type of offer, the character of the work and the Conditions of Contract. Where a standard Tender Form accompanies standard Conditions then that form (with amendments,where necessary) will obviously be adopted, but otherwise a form will have to be drafted; the items to be included, in their logical order, are:

(1) a recital setting out the documents (both Employer's and Tenderer's documents) forming part of the tender and confirming that those issued by the Employer had been seen by the tenderer;

(2) an offer to perform the work set out in the documents for a particular sum or sums; alternatively, to carry out the work for a sum or sums calculated in accordance with the provisions of a named document such as, for example, "the Schedule of Rates" or "the Bills of Quantities";

(3) a statement that the tenderer will complete the work in a specific time or times; alternatively, that the tenderer will complete the work in a time or times set out in named document, such as 'the Appendix hereto';

(4) a statement agreeing that the tender will remain open for a stated period from the time for submission;

(5) a statement agreeing that until such time as a formal agreement is entered into between the Contractor and the Employer, the Tender and the Employer's written acceptance thereof shall form a binding contract;

(6) when the Conditions of Contract require it and it is to form part of the contract, a statement that the tenderer will provide a Performance Bond (or a Guarantee) for an amount equal to a stated percentage of the Tender sum within a stated period of receiving a request for it from the Employer.

When tenders are called by public authorities it is usual to add a statement that the tenderer understands that the Employer is not bound to accept the lowest or any tender. It is not necessary for this to be included, but it is usually considered desirable to assist in avoiding disputes if the Authority does not accept the lowest tenderer.

Some Conditions of Contract include a condition that the Contractor shall provide a performance bond or guarantee 'if this requirement is included in the Tender'. The provision in the Tender for this bond is, therefore, only required when there is a corresponding condition in the Conditions of Contract and when the Employer wishes to have a performance bond. As pointed out above, many public authorities find that they rarely have to call upon the performance bond and it is, therefore, more expensive to pay for a bond on all contracts rather than to meet the additional cost of the rare contract which involves contractor's failure.

There has been considerable discussion in the UK concerning the possibility or the desirability of holding a tenderer to his tender for the period of tender validity. The tender offer is not a contract until it is accepted and, therefore, it is not possible to enforce any term, such as the validity period. It is also contended that contract arrangements are likely to be unsatisfactory if the contract is made with an unwilling contractor and that, for this reason, a tenderer should not be forced into a contract which he realises would be to his severe disadvantage. Most tenderers, of course, are only too willing to have their tenders accepted! For these reasons many U.K. tender forms do not include a term concerning period of tender validity. However, it is usual to provide a tender validity period in international contracts and, in order to enforce this, the tenderer will be required to either deposit a sum of money (usually referred to as Earnest Money) or (more commonly today) deposit a bond for a stated sum from a bank or insurance company, the bond being under seal where this is required by the law of the Employer's country. Requirements concerning Earnest Money or Tender Bond are not usually included in the Tender Form, but tenderers are instructed to provide it and informed that their tender will not be considered unless it is provided.

Where an Appendix is attached to the Tender Form (in order to give specific information not included elsewhere in the tender documents) then it is essential to say, on the Tender Form, that "The Appendix forms part of the Tender". The following list forms a useful aide mémoire of items which may be required in the Appendix:—

(1) Completion periods and liquidated damages per day or week in respect of each of the completion periods.
(2) Limit of liquidated damages (where this is to be included).
(3) Particulars of bonuses, if applicable.
(4) The Maintenance Periods or the Periods of Defects Liability

(there are sometimes different periods for different parts of the Works).

(5) Minimum amount of third party insurance.

(6) Amount of Performance Bond or Guarantee (if any) as percentage of Tender Sum.

(7) Percentage for adjustment of P.C. sums.

(8) Percentage of the value of material to be included in interim certificates.

(9) Particulars of materials which are to be stored off site but whose value is to be included in interim certificates.

(10) Percentage of retention.

(11) Limit of retention.

(12) Minimum amount of interim certificates.

(13) Time for payment after issue of certificate.

It is unlikely that all the foregoing items will be required for inclusion in the Appendix; some contracts may require other items which are not of general interest such as the Method of Measurement used in an accompanying Bill of Quantities. Generally, the value or percentages in the various items are inserted by the Employer or his consulting engineer, but the item of 'Percentage for adjustment of P.C. sums' has to be inserted by the tenderer; in many tenders the tenderer is also required to insert the time or times for completion. The items to be inserted by the tenderer should be clearly marked (usually with an asterisk) and a note included in the Appendix stating that the tenderer is to insert the figures in those items.

7.3 Acceptance and Formal Agreement

When an acceptable offer has been received by the Employer, whether by competitive tender or by negotiation, then a formal acceptance of the offer must be made if he wishes work to proceed. In the UK this is usually in the form of a simple letter of acceptance which often provides for a formal Agreement to follow, but in some other countries (particularly where Earnest Money or a Tender Bond is provided) the acceptance may be in the form of an instruction to the tendering contractor to send his authorised representative to sign a formal agreement at a specified date and time. A typical UK acceptance letter might read –

We are pleased to inform you that we have decided to accept your

Tender dated ....*[or 'the offer set out in your letter dated .....']* for
the construction of .... A formal contract Agreement will be
prepared as soon as possible, but meanwhile your tender *[or 'offer']*
and this letter of acceptance will constitute a binding contract between
us.

If the Employer requires a performance bond, a further sentence should
be added instructing the tenderer to supply such a bond within a stated
period, say fourteen or twenty one days.

The main advantage of a formal agreement in engineering contracts
is that it specifies and lists the documents of the contract and thus excludes
other documents which may have been introduced into negotiations or
which (such as "Instructions to Tenderers") are not intended to form
part of the contract. It thus fully defines the extent of the contract. If
the Agreement is in the form of a deed then there is the further advantage
that the limitation period is extended from six years to twelve years.

Engineers are not often required to draft formal agreements but will
usually have to tailor a standard agreement form to the requirements of
the Contract. A form of Agreement for an engineering contract will
consist of the following items:—

(1) A statement of the date and of the titles and addresses of the
parties.
(2) A recital commencing with the words "Whereas..." and stating
that the Employer desires that the ......*[title of the Works]* be
executed and has accepted the Contractors Tender.
(3) A statement of the substance of the contract commencing with
the words "NOW THIS AGREEMENT PROVIDES as
follows :".
(4) The signing clause, commencing "IN WITNESS whereof the
parties hereto have caused this Agreement to be signed by their
duly authorised representatives on the date first above written.",
followed by lines for the signature and the names of the parties
and for witnesses.

Item 3 of the Agreement will usually consist of four clauses as
follows:—

(i) A clause stating that in the Agreement words are as defined in
the Conditions of Contract.
(ii) A list of the documents, preceded by introductory wording such

as "The following documents shall be read and construed as part of this Agreement:".

(iii) A statement that, in consideration of the payments made by the Employer the Contractor covenants to execute the Works in accordance with the Contract.

(iv) A statement that the Employer covenants to pay the Contractor the prices prescribed by the Contract for the execution of the Works.

The precise wording of the Agreement will depend upon the requirements of the contract; in many, the wording adopted in standard forms can be used, modified to suit the actual requirements. Although some standard forms (such as the I.Chem.E. Agreement form) may appear to vary in some respects, it will be found that they all contain the items listed above together with modifications and additions to suit the circumstances for which they were drafted.

7.4 Non-contractual Documents

Prior to the formation of a contract it is usually necessary to provide certain documents which are related to the contract but which do not form part of it. In the case of a negotiated contract there will be correspondence and preliminary documents which are subsequently amended or superseded. The principles of drafting apply equally to the preliminary documents. When the contract is based upon a competitive tender, a number of documents are usually issued which are concerned with the process of tendering and which do not form part of the contract; in the case of public authorities these tender documents are often of a semi-standard nature.

7.4.1 Pre-qualification

In the past it was common for public authorities to call open tenders, i.e. they would advertise and allow anyone to tender for the advertised contract. In many cases this resulted in unsatisfactory tenders or in the work being carried out by incompetent or financially unsound contractors. The practice today is for public authorities to maintain lists of contractors who are qualified for particular types and sizes of contract and allow tenders only from contractors on the list or, in the case of large contracts, from a small number of contractors specially selected from those on the appropriate list. In order to be included in an authority's list of qualified contractors, an applicant must provide the information set out in a standard questionnaire issued by the Authority and may be subject to further enquiries as a result of the answers given. Engineers are sometimes required to prepare such questionnaires for a client; the questions are usually based on the following items, but they may vary according to the class of work and the type of contract:—

(1) Type of work and size and type of contract for which pre-qualification is sought.
(2) Corporate particulars of the applicant, such as title and registered address, particulars of company registration, of capital, of whether a member of a group and whether there are any subsidiary companies, particulars of management control, etc.
(3) Classes of work carried out by the company and particulars of relevant recent contracts and references from clients.
(4) Particulars of staff and other employees (curriculum vitae of senior staff to be included) and of any consultants regularly employed; in some cases an organisation chart may be called for.
(5) Lists of plant and/or manufacturing facilities owned.
(6) Audited financial accounts, usually for the past three years.

Many authorities have special requirements and the actual questions will need to be drafted in relation to these and the policy of the authority concerned. Overseas authorities often have special local requirements such as registration in the country concerned, acceptability of bankers, etc.

7.4.2 Invitation and Instructions to Tenderers

When tender documents are issued they are usually sent under cover of a formal "Invitation to Tender". This may be in the form of a simple letter listing the documents which are enclosed with it and giving the tender date. It is more usual, however, to give fuller particulars even though these may repeat information given in the "Instructions to Tenderers". The information given in the Invitation to Tender is usually limited to the following:—

(a) An introductory paragraph describing briefly the type and extent of the work.

(b) The type of contract, e.g. lump sum or remeasurement and whether manufacture and erection or construction only or design and build.

(c) Tender date.

(d) The period for which the tender is to remain valid after the tender date and whether earnest money or a tender bond is required.

The tender documents and the Invitation to Tender are invariably accompanied by a document entitled "Instructions to Tenderers" or sometimes "Instructions for Tendering". These Instructions do not normally form part of the contract, although there are a few public authorities (particularly overseas) who do require the Instructions to form part of the contract. The Instructions to Tenderers are intended to set out the detailed tendering requirements and usually include the following items (set out in logical order):—

(1) An introductory paragraph giving the title of the Works and a very brief description of the type of work concerned.

(2) Instructions concerning submission of the tender, dealing with:
  (i) Completion of the relevant parts of the documents (typically "Tenderers shall complete the Form of Tender, and each separate item in the Schedule of Prices clearly and in ink");
  (ii) the type of envelope or envelopes in which the tender is to be submitted;
  (iii) the address to which the tender is to be sent together with the date and time by which the Tender is to be submitted (to allow for any delay in completion and approval of the documents it is common to give the date in the Invitation to Tender and to merely refer to that date in the Instruction);
  (iv) when all the documents which are to be submitted are not bound in one volume, a list of the documents which are to accompany the Tender;
  (v) requirements for return of documents not returned with the Tender, together with the address to which they are to be returned; this address will not usually be the same as the address for submission of the Tender (although it is sometimes merely a different room in the same building) but is commonly the address of a consulting engineer who has prepared the documents).

(3) A statement that tenders shall be submitted strictly in accordance

with the tender documents and that failure to comply with this requirement may invalidate the tender.

(4) When appropriate, instructions concerning the submission of alternative proposals, including the following matters:

    (i) alternative proposals will only be considered if a tender is submitted in accordance with the Tender Documents;

    (ii) alternative proposals shall be in the form of a tender with fully priced Schedule or Bill and sufficient drawings and explanatory matter to enable them to be technically assessed;

    (iii) when alternative proposals are for a substantially different design, the tenderer may be permitted to discuss his proposal with the consulting engineers or designers to enable the Tender to meet the general design requirements.

(5) Provisions to enable tenderers to raise queries concerning the documents; these must be in writing; replies will be circulated to all tenderers.

(6) Provisions for the client to amend the documents during the tendering period and also to extend the tender period, if required.

(7) Particulars of any reports or other documents to be made available to tenderers, either for inspection at a specified address or by copies issued with the tender document (typically – design reports, site investigation reports, and the like).

(8) Particulars of documents which are to accompany the Tender (such as preliminary programmes and method statements showing the proposed manner of carrying out the work) which are intended for guidance only and are to be non-contractual.

(9) Paragraphs drawing attention to particular points in the tender documents, such as whether there is provision for variation of the price to deal with inflation, any provision for an initial advance for moblisation, whether alternatives dates for completion are to be offered, any requirements for performance bond and any other bonding (such as for retention), any applicable legislation which must be complied with, etc.

(10) Arrangements for announcing the results of tendering.

(11) Paragraphs stating that the cost of tendering is to be entirely borne by the tenderer and that the Employer does not bind himself to accept the lowest or any tender.

Although these items are those usually included in the UK, many international authorities (e.g. the World Bank) require the inclusion of a wide variety of other items such as the source of the funds for carrying out the work, the languages of the tender, the currencies of the tender and of payments under the contract, the legal authority of the person signing the tender, the procedure for tender opening, the procedure which will be adopted for evaluation and comparison of tenders, the procedure for award of contract, etc. As mentioned earlier, many Authorities have model forms of Instructions to Tenderers and the draftsman will be bound to follow the layout, requirements and detailed drafting of the model, but will have to add to and vary it to suit the particular contract procedures and documents. Although the Instructions to Tenderers does not usually form part of a contract, it must not contain matter which might be misleading to tenderers and which could, therefore, lead to claims of misrepresentation.

Bibliography

1 Writing

Engineering draftsmen are advised to read the following two books to assist in writing clear English:—

The Complete Plain Words by Sir Ernest Gowers. Her Majesty's Stationery Office and Pelican Books. 1962.

Mind the Stop by G. V. Carey. Cambridge University Press. 1939.

Although the latter book is devoted exclusively to the use of punctuation, it does not deal with the punctuation of marshalled sentences for document drafting.

2 Legal Drafting

Books on the general aspects of legal drafting, as distinct from the drafting of specific legal documents, are few and far between. Only three have been identified as of interest to engineers.

The Elements of Drafting by E. L. Piesse and J. H. Aitken. The Law Book Company Ltd. 1981.

This is a short book with a concise text which covers most of the aspects that engineers are likely to require.

The Fundamentals of Legal Drafting by Reed Dickerson. Little, Brown and Company. 1986.

This was written at the request of the American Bar Association. It includes philosophical and literary comments on the use of language.

Drafting by Stanley Robinson. Butterworth and Company (Publishers) Ltd. 1980.

The first half of this book covers the same ground as Piesse and Aitken. The second half deals with the drafting of specific legal documents.

3 Conditions of Contract

The following list covers most of the published standard forms in use in the UK for domestic and international contracts.

(i) UK Government Contracts

Form GC/Works/1. General Conditions of Government Contracts for Building and Civil Engineering Works. 1977.

Form GC/Works/2. General Conditions of Government Contracts for Building and Civil Engineering Minor Works. 1980.

Form C1001. General Conditions of Government Contracts for Building, Civil Engineering, Mechanical and Electrical Small Works. 1982.

All are published by Her Majesty's Stationery Office. They are used mainly by the Property Services Agency; other government offices often use one of the standard forms listed below.

(ii) Civil Engineering

Conditions of Contract and Forms of Tender, Agreement and Bond for use in connection with Works of Civil Engineering Construction. (ICE Conditions of Contract). Fifth edition as amended 1986.

Conditions of Contract, Agreement and Contract Schedule for use in connection with Minor Works in Civil Engineering Construction. 1988.

Conditions of Contract and Forms of Tender, Agreement and Bond for use in connection with Ground Investigation. 1983.

These 3 are published by the Institution of Civil Engineers (ICE), the Association of Consulting Engineers (ACE) and the Federation of Civil Engineering Contractors (FCEC).

Conditions of Contract for Overseas Works Mainly of Civil Engineering Construction with Forms of Tender and Agreement. The ACE, the ICE, and the Export Group for the Construction Industries. 1956.

FIDIC Conditions of Contract for Works of Civil Engineering Construction. Part I General Conditions with Forms of Tender and Agreement. Part II Conditions of Particular Application with Guidelines for Preparation of Part II Clauses. Fédération Internationale des Ingénieurs-Conseils. 1987.

(iii) Mechanical and Electrical Engineering

Model Form of General Conditions of Contract including Forms of Agreement and Guarantee recommended by the Institution of Mechanical Engineers, the Institution of Electrical Engineers and the Association of Consulting Engineers for use in connection with Home Contracts — With Erection. (Model Form A). Institution of Electrical Engineers (IEE). 1982.

Model Form of General Conditions of Contract including Form of Agreement recommended by the Institution of Mechanical Engineers, the Institution of Electrical Engineers and the Association of Consulting Engineers for use in connection with Export Contracts for Supply of Plant and Machinery (Electrical & Mechanical). (Model Form B1). IEE. Fifth Edition 1981.

Model Form of General Conditions of Contract including Form of Agreement recommended by the Institution of Mechanical Engineers, the Institution of Electrical Engineers and the Association of Consulting Engineers for use in connection with Export Contracts. Delivery FOB or CIF or FOR with Supervision of Erection. (Electrical & Mechanical). (Model Form B2). IEE. Fifth Edition 1981.

Model Form of General Conditions of Contract including Form of Agreement recommended by the Institution of Mechanical Engineers, the Institution of Electrical Engineers and the Association of Consulting Engineers for use in connection with Export Contracts (including Delivery to and Erection on Site of Electrical and Mechanical & Mechanical Plant). (Model Form B3). IEE. Third Edition 1980.

Model Form of General Conditions of Contract recommended by the Institution of Mechanical Engineers and the Institution of Electrical Engineers for the supply of Electrical and Mechanical Goods, other than Electric Cables (Home — without erection). (Model Form C). IEE. 1975.

Model Form of General Conditions of Contract recommended by the Institution of Electrical Engineers for Home Cable Contracts with Installation. (Model Form E). IEE. 1982.

FIDIC Conditions of Contract for Electrical & Mechanical Works (including Erection on Site) with Forms of Tender and Agreement. Fédération Internationale des Ingénieurs–Conseil. 1987.

(iv) Chemical Engineering

Model Form of Conditions of Contract for Process Plants Suitable for reimbursible contracts in the United Kingdom. Institution of Chemical Engineers (I.Chem.E). 1976.

Suitable for lump-sum contracts in the United Kingdom. I.Chem.E. 1981.

(v) Building

Standard Form of Building Contract. JCT80.
— *Local Authorities with Quantities.* 1986.
— *Local Authorities without Quantities.* 1986.
— *Local Authorities with Approximate Quantities.* 1986.
— *Private with Quantities.* 1986.
— *Private with Approximate Quantities.* 1986.
— *Private without Quantities.* 1986.

Standard Form With Contractors Design. CD81. 1986.

Intermediate Form of Building Contract. IFC84. 1986.

Agreement for Minor Works. 1987.

Fixed fee form of prime cost contract. 1976.

Agreement for renovation grant works where architect is employed. 1974.

Agreement for renovation grant works where architect is not employed. 1974.

All published for the Joint Contracts Tribunal (JCT) by RIBA Publications Ltd.

ACA Form of Building Agreement (2nd. Edition). Association of Consulting Architects. 1984.

BP88 Form of Building Agreement. British Property Federation. 1984.

4 Conditions of Sub-contract

These published subcontract conditions are drafted for use with those Conditions of Contract (listed above) to which they refer.

Form of Sub-contract designed for use in conjunction with the ICE General Conditions of Contract. The Federation of Civil Engineering Contractors. 1984.

JCT80. Nominated Sub-Contract for sub-contractors who have tendered. NSC/4.

JCT80. Nominated Sub-Contract for sub-contractors who have not tendered. NSC/4a.

Both published for JCT by RIBA Publications Ltd. 1986.

Domestic Sub-Contract Conditions for use with JCT80. DOM/1c.

Conditions for domestic sub-contracts under IFC84. IN/SC.

Domestic Sub-Contract, Articles for use with Contractor's Design 1981. DOM/2.

These three by Building Employers Confereration. 1986.

5 Standard Specifications

The following are published standard specifications drafted for use with the Conditions of Contract stated. They incorporate the necessary contractual provisions to enable them to form part of a contract which is subject to the stated Conditions, but require modification if they are to be used with other Conditions. They are not comparable to British Standard Specifications, which do not include contractual provisions.

Specification for Highway Works. For use with ICE Conditions. Her Majesty's Stationery Office. 1986.

Civil engineering specification for the water industry. For use with ICE Conditions. Water Research Association. 1984.

Building and Civil Engineering Minor Works. For use with Form GC/Works/2. Property Services Agency. 1986.

Index